Supercharge Your Applications with GraalVM

Hands-on examples to optimize and extend your code using GraalVM's high performance and polyglot capabilities

A B Vijay Kumar

BIRMINGHAM—MUMBAI

Supercharge Your Applications with GraalVM

Copyright © 2021 Packt Publishing

Group Product Manager: Aaron Lazar
Publishing Product Manager: Kushal Dave
Senior Editor: Rohit Singh
Content Development Editor: Kinnari Chohan
Technical Editor: Karan Solanki
Copy Editor: Safis Editing
Project Coordinator: Francy Puthiry
Proofreader: Safis Editing
Indexer: Vinayak Purushotham
Production Designer: Shankar Kalbhor

First published: May 2021

Production reference: 3230821

Published by Packt Publishing Ltd.
Livery Place
35 Livery Street
Birmingham
B3 2PB, UK.

ISBN 978-1-80056-490-9

www.packt.com

To the memory of my father, Amba Prasad, and my mother, Vasantha, for all their sacrifice and upbringing. I miss you both, but I am sure you are always around me, guiding me.

To my wife, Geetha, for all her love, support, and encouragement. Without her, I wouldn't have come this far. And to my new love, Bozo, our puppy.

Contributors

About the author

A B Vijay Kumar is an "IBM Distinguished Engineer" and chief technology officer focused on hybrid cloud management and platform engineering. He is responsible for providing technology strategies for managing complex application portfolios on hybrid cloud platforms using emerging tools and technologies.

He is an IBM Master Inventor who has more than 31 patents issued and 30 pending in his name. He has more than 23 years' experience at IBM. He is recognized as a subject matter expert for his contribution to advanced mobility in automation and has led several implementations involving complex industry solutions. He specializes in mobile, cloud, container, automotive, sensor-based machine-to-machine, Internet of Things, and telematics technologies.

I wish to thank my family, especially my lovely brothers, Ram and Shyam, and my extended family – Saroja, Priya Shyam, Suresh, Priya Balu, Kuvalesh, Ritvik, Rijjul, Midhushi, and other family members, for all their support and encouragement.

I wish to thank my colleagues and friends, Jhilam B, Joyil J, Naveen E, Archan G, Amit D, Arun N, and Vasu R, who have provided critical feedback and helped me through the journey of writing this book.

Special thanks to Chris Seaton, from Shopify, for helping me understand some tough concepts, and debugging my issues. I would also like to thank IBM Corp and my management for encouraging me and allowing me to write this book.

Last but not least, thanks to the awesome Packt team – Kunal, Rohit, Kinnari, Prajakta, Kushal, Karan, and everybody else behind the scenes, for their awesome support, without which this book would never have materialized.

About the reviewer

Esteban Ginez, a seasoned developer, currently works at the intersection of cloud infrastructure, web services, and new compiler tooling.

During his tenure at Oracle, he spent his time improving how people use web services. During his time on the GraalVM team, he worked on features making Java and the JVM better suited to cloud workloads. In the past, Esteban has worked at a variety of tech companies, including Amazon and Zillow. In his spare time, Esteban enjoys contributing to open source projects. Originally from Quito, Ecuador, Esteban graduated from the University of Calgary with a degree in computer science.

Table of Contents

Section 2: Getting Up and Running with GraalVM – Architecture and Implementation

3

GraalVM Architecture

4
Graal Just-In-Time Compiler

5
Graal Ahead-of-Time Compiler and Native Image

Section 3: Polyglot with Graal

6

Truffle for Multi-language (Polyglot) support

7

GraalVM Polyglot – JavaScript and Node.js

8
GraalVM Polyglot – Java on Truffle, Python, and R

9
GraalVM Polyglot – LLVM, Ruby, and WASM

Section 4: Microservices with Graal

10

Microservices Architecture with GraalVM

Assessments

Other Books You May Enjoy

Index

Preface

GraalVM is a universal virtual machine that allows programmers to embed, compile, interoperate, and run applications written in JVM languages such as Java, Kotlin, and Groovy; non-JVM languages such as JavaScript, Python, WebAssembly, Ruby, and R; and LLVM languages such as C and C++.

GraalVM provides the Graal **just-in-time** (**JIT**) compiler, an implementation of the Java Virtual Machine Compiler Interface (**JVMCI**), which is completely built on Java and uses Java JIT compiler (C2 compiler) optimization techniques as the baseline and builds on top of them. The Graal JIT compiler is much more sophisticated than the Java C2 JIT compiler. GraalVM is a drop-in replacement for the JDK, which means that all the applications that are currently running on the JDK should run on GraalVM without any application code changes.

GraalVM also provides **ahead-of-time** (**AOT**) compilation to build native images with static linking. GraalVM AOT compilation helps build native images that have a very small footprint and faster startup and execution, which is ideal for modern-day microservices architectures.

While GraalVM is built on Java, it not only supports Java, but also enables polyglot development with JavaScript, Python, R, Ruby, C, and C++. It provides an extensible framework called Truffle that allows any language to be built and run on the platform.

GraalVM is becoming the default runtime for running cloud-native Java microservices. Soon, all Java developers will be using GraalVM to run their cloud-native Java microservices. There are already a lot of microservices frameworks that are emerging in the market, such as Quarkus, Micronaut, Spring Native, and so on, that are built on GraalVM.

Developers working with Java will be able to put their knowledge to work with this practical guide to GraalVM and cloud-native microservice Java frameworks. The book provides a hands-on approach to implementation and associated methodologies that will have you up and running, and productive in no time. The book also provides step-by-step explanations of essential concepts with simple and easy-to-understand examples.

This book is a hands-on guide for developers who wish to optimize their apps' performance and are looking for solutions. We will start by giving a quick introduction to the GraalVM architecture and how things work under the hood. Developers will quickly move on to explore the performance benefits they can gain by running their Java applications on GraalVM. We'll learn how to create native images and understand how AOT can improve application performance significantly. We'll then move on to explore examples of building polyglot applications and explore the interoperability between languages running on the same VM. We'll explore the Truffle framework to implement our own languages to run optimally on GraalVM. Finally, we'll also learn how GraalVM is specifically beneficial in cloud-native and microservices development.

Who this book is for

The primary audience for this book is JVM developers looking to optimize their application's performance. This book would also be useful to JVM developers who are exploring options to develop polyglot applications by using tooling from the Python/R/Ruby/Node.js ecosystem. Since this book is for experienced developers/programmers, readers must be well-versed in software development concepts and should have good knowledge of working with programming languages.

What this book covers

Chapter 1, *Evolution of Java Virtual Machine*, walks through the evolution of JVM and how it optimized the interpreter and compiler. It will walk through C1 and C2 compilers, and the kind of code optimizations that JVM performs to run Java programs faster.

Chapter 2, *JIT, HotSpot, and GraalJIT*, takes a deep dive into how JIT compilers and Java HotSpot work and how JVM optimizes code at runtime.

Chapter 3, *GraalVM Architecture*, provides an architecture overview of Graal and the various architecture components. The chapter goes into details on how GraalVM works and how it provides a single VM for multiple language implementations. This chapter also covers the optimizations GraalVM brings on top of standard JVM.

Chapter 4, *Graal Just-In-Time Compiler*, talks about the JIT compilation option of GraalVM. It goes through the various optimizations the Graal JIT compiler performs in detail. This is followed by a hands-on tutorial to use various compiler options to optimize the execution.

Chapter 5, *Graal Ahead-of-Time Compiler and Native Image*, is a hands-on tutorial that walks through how to build native images and optimize and run these images with profile-guided optimization techniques.

Chapter 6, Truffle for Multi-language (Polyglot) support, introduces the Truffle polyglot interoperability capabilities and high-level framework components. It also covers how data can be transferred between applications that are written in different languages that are running on GraalVM.

Chapter 7, GraalVM Polyglot – JavaScript and Node.js, introduces the JavaScript and NodeJs. This is followed by a tutorial on how to use the Polyglot API for interoperability to interoperate between an example JavaScript and NodeJS application and a Python application.

Chapter 8, GraalVM Polyglot – Java on Truffle, Python, and R, introduces Python, R, and Java on Truffle (Espresso). This is followed by a tutorial on how to use the Polyglot API for interoperability between various languages.

Chapter 9, GraalVM Polyglot – LLVM, Ruby, and WASM, introduces JavaScript and Node.js. This is followed by a tutorial on how to use the Polyglot API to interoperate between an example JavaScript/Node.js applications.

Chapter 10, Microservices Architecture with GraalVM, covers the modern microservices architecture and how new frameworks such as Quarkus and Micronaut implement Graal for the most optimum microservices architecture.

To get the most out of this book

This book is a hands-on guide, with step-by-step instructions on how to work with GraalVM. Throughout the book, the author has used very simple, easy-to-understand code samples that will help you to understand the core concepts of GraalVM. All the code samples are provided in a Git repository. You are expected to have good knowledge of the Java programming language. The book also touches upon Python, JavaScript, Node.js, Ruby, and R – but the examples are intentionally kept simple, for understanding, while focusing on demonstrating the polyglot interoperability concepts.

Software/Hardware covered in the book	OS Requirements
GraalVM (preferably the Enterprise Edition – which comes with all the runtimes and tools that are required)	Windows, macOS, and Linux (any)
Visual Studio Code (or any other IDE)	
Git	
Maven	
OpenJDK	

If you are using the digital version of this book, we advise you to type the code yourself or access the code via the GitHub repository (link available in the next section). Doing so will help you avoid any potential errors related to the copying and pasting of code.

Download the example code files

You can download the example code files for this book from GitHub at `https://github.com/PacktPublishing/Supercharge-Your-Applications-with-GraalVM`. In case there's an update to the code, it will be updated on the existing GitHub repository.

We also have other code bundles from our rich catalog of books and videos available at `https://github.com/PacktPublishing/`. Check them out!

Code in Action

Code in Action videos for this book can be viewed at `https://bit.ly/3eM5ewO`.

Download the color images

We also provide a PDF file that has color images of the screenshots/diagrams used in this book. You can download it here: `https://static.packt-cdn.com/downloads/9781800564909_ColorImages.pdf`.

Conventions used

There are a number of text conventions used throughout this book.

`Code in text`: Indicates code words in text, database table names, folder names, filenames, file extensions, pathnames, dummy URLs, user input, and Twitter handles. Here is an example: "In Truffle, it is a Java class derived from `com.oracle.truffle.api.nodes.Node`."

A block of code is set as follows:

```
@Fallback protected void typeError (Object left, Object right)
{
    throw new TypeException("type error: args must be two
        integers or floats or two", this);
}
```

Any command-line input or output is written as follows:

```
X /Library/Java/JavaVirtualMachines/graalvm-ee-
java11-21.0.0.2/Contents/Home/bin/npm --version
6.14.10
```

Bold: Indicates a new term, an important word, or words that you see onscreen. For example, words in menus or dialog boxes appear in the text like this. Here is an example: "Select **System info** from the **Administration** panel."

Tips or important notes
Appear like this.

Get in touch

Feedback from our readers is always welcome.

General feedback: If you have questions about any aspect of this book, mention the book title in the subject of your message and email us at customercare@packtpub.com.

Errata: Although we have taken every care to ensure the accuracy of our content, mistakes do happen. If you have found a mistake in this book, we would be grateful if you would report this to us. Please visit www.packtpub.com/support/errata, selecting your book, clicking on the Errata Submission Form link, and entering the details.

Piracy: If you come across any illegal copies of our works in any form on the Internet, we would be grateful if you would provide us with the location address or website name. Please contact us at copyright@packt.com with a link to the material.

If you are interested in becoming an author: If there is a topic that you have expertise in and you are interested in either writing or contributing to a book, please visit authors.packtpub.com.

Reviews

Please leave a review. Once you have read and used this book, why not leave a review on the site that you purchased it from? Potential readers can then see and use your unbiased opinion to make purchase decisions, we at Packt can understand what you think about our products, and our authors can see your feedback on their book. Thank you!

For more information about Packt, please visit packt.com.

Section 1:
The Evolution
of JVM

This section will walk through the evolution of JVM, and how it optimized the interpreter and compiler. It will walk through C1 and C2 compilers, and the kind of code optimizations that JVM performs to run Java programs faster.

This section comprises the following chapters:

- *Chapter 1, Evolution of Java Virtual Machine*
- *Chapter 2, JIT, HotSpot, and GraalJIT*

1
Evolution of Java Virtual Machine

This chapter will walk you through the evolution of **Java Virtual Machine (JVM)**, and how it optimized the interpreter and compiler. We will learn about C1 and C2 compilers and various types of code optimizations that the JVM performs to run Java programs faster.

In this chapter, we will cover the following topics:

- Introduction to GraalVM
- Learning how JVM works
- Understanding the JVM architecture
- Understanding the kind of optimizations JVM performs with **Just-In-Time (JIT)** compilers
- Learning the pros and cons of the JVM approach

By the end of this chapter, you will have a clear understanding of the JVM architecture. This is critical in understanding the GraalVM architecture and how GraalVM further optimizes and builds on top of JVM best practices.

Technical requirements

This chapter does not require any specific software/hardware.

Introduction to GraalVM

GraalVM is a high-performance VM that provides the runtime for modern cloud-native applications. Cloud-native applications are built based on the service architecture. The microservice architecture changes the paradigm of building micro applications, which challenges the fundamental way we build and run applications. The microservices runtimes demand a different set of requirements.

Here are some of the key requirements of a cloud-native application built on the microservice architecture:

- **Smaller footprint**: Cloud-native applications run on the "pay for what we use" model. This means that the cloud-native runtimes need to have a smaller memory footprint and should run with the optimum CPU cycles. This will help run more workloads with fewer cloud resources.

- **Quicker bootstrap**: Scalability is one of the most important aspects of container-based microservices architecture. The faster the application's bootup, the faster it can scale the clusters. This is even more important for serverless architectures, where the code is initialized and run and then shut down on request.

- **Polyglot and interoperability**: Polyglot is the reality; each language has its strengths and will continue to. Cloud-native microservices are being built with different languages. It's very important to have an architecture that embraces the polyglot requirements and provides interoperability across languages. As we move to modern architectures, it's important to reuse as much code and logic as possible, that is time-tested and critical for business.

GraalVM provides a solution to all these requirements and provides a common platform to embed and run polyglot cloud-native applications. It is built on JVM and brings in further optimizations. Before understanding how GraalVM works, it's important to understand the internal workings of JVM.

Traditional JVM (before GraalVM) has evolved into the most mature runtime implementation. While it has some of the previously listed requirements, it is not built for cloud-native applications, and it comes with its baggage of monolith design principles. It is not an ideal runtime for cloud-native applications.

This chapter will walk you through in detail how JVM works and the key components of the JVM architecture.

Learning how JVM works

Java is one of the most successful and widely used languages. Java has been very successful because of its *write once, run anywhere* design principle. JVM realizes this design principle by sitting between the application code and the machine code and interpreting the application code to machine code.

Traditionally, there two ways of running application code:

- **Compilers**: Application code is directly compiled to machine code (in C, C++). Compilers go through a build process of converting the application code to machine code. Compilers generate the most optimized code for a specific target architecture. The application code has to be compiled to target architectures. In general, the compiled code always runs faster than interpreted code, and issues with code semantics can be identified during compilation time rather than runtime.

- **Interpreters**: Application code is interpreted to machine code line by line (JavaScript and so on). Since interpreters run line by line, the code may not be optimized to the target architecture, and run slowly, compared to the compiled code. Interpreters have the flexibility of writing once and running anywhere. A good example is the JavaScript code that is predominantly used for web applications. This runs pretty much on different target browsers with minimal or no changes in the application code. Interpreters are generally slow and are good for running small applications.

JVM has taken the best of both interpreters and compilers. The following diagram illustrates how JVM runs the Java code using both the interpreter and compiler approaches:

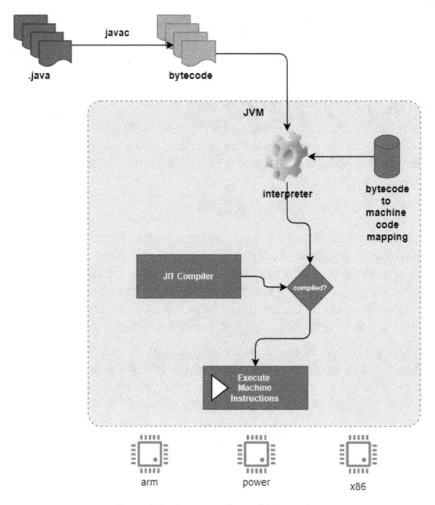

Figure 1.1 – Java compiler and interpreter

Let's see how this works:

- Java Compiler (**javac**) compiles the Java application source code to **bytecode** (intermediate format).

- JVM interprets the bytecode to machine code line by line at runtime. This helps in translating the optimized bytecode to target machine code, helping in running the same application code on different target machines, without re-programming or re-compiling.

- JVM also has a Just-In-Time (**JIT**) compiler to further optimize the code at runtime by profiling the code.

In this section, we looked at how Java Compiler and JIT work together to run Java code on JVM at a higher level. In the next section, we will learn about the architecture of JVM.

Understanding the JVM architecture

Over the years, JVM has evolved into the most mature VM runtime. It has a very structured and sophisticated implementation of a runtime. This is one of the reasons why GraalVM is built to utilize all the best features of the JVM and provide further optimizations required for the cloud-native world. To better appreciate the GraalVM architecture and optimizations that it brings on top of the JVM, it's important to understand the JVM architecture.

This section walks you through the JVM architecture in detail. The following diagram shows the high-level architecture of various subsystems in JVM:

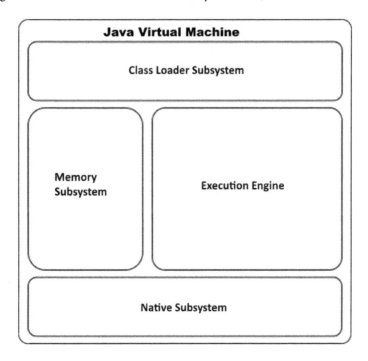

Figure 1.2 – High-level architecture of JVM

The rest of this section will walk you through each of these subsystems in detail.

Class loader subsystem

The class loader subsystem is responsible for allocating all the relevant .class files and loading these classes to the memory. The class loader subsystem is also responsible for linking and verifying the schematics of the .class file before the classes are initialized and loaded to memory. The class loader subsystem has the following three key functionalities:

- Loading
- Linking
- Initializing

The following diagram shows the various components of the class loader subsystem:

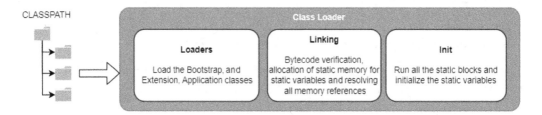

Figure 1.3 – Components of the class loader subsystem

Let's now look at what each of these components does.

Loading

In traditional compiler-based languages such as C/C++, the source code is compiled to object code, and then all the dependent object code is linked by a linker before the final executable is built. All this is part of the build process. Once the final executable is built, it is then loaded into the memory by the loader. Java works differently.

Java source code (`.java`) is compiled by Java Compiler (`javac`) to bytecode (`.class`) files. Class loader is one of the key subsystems of the JVM, which is responsible for loading all the dependent classes that are required to run the application. This includes the classes that are written by the application developer, the libraries, and the **Java Software Development Kit (SDK)** classes.

There are three types of class loaders as part of this system:

- **Bootstrap**: Bootstrap is the first classloader that loads `rt.jar`, which contains all the Java Standard Edition JDK classes, such as `java.lang`, `java.net`, `java.util`, and `java.io`. Bootstrap is responsible for loading all the classes that are required to run any Java application. This is a core part of the JVM and is implemented in the native language.

- **Extensions**: Extension class loaders load all the extensions to the JDK found in the `jre/lib/ext` directory. Extension class loader classes are typically extension classes of the bootstrap implemented in Java. The extension class loader is implemented in Java (`sun.misc.Launcher$ExtClassLoader.class`).

- **Application**: The application class loader (also referred to as a system class loader) is a child class of the extension class loader. The application class loader is responsible for loading the application classes in the application class path (`CLASSPATH` env variable). This is also implemented in Java (`sun.misc.Launcher$AppClassLoader.class`).

Bootstrap, extension, and application class loaders are responsible for loading all the classes that are required to run the application. In the event where the class loaders do not find the required classes, `ClassNotFoundException` is thrown.

Class loaders implement the delegation hierarchy algorithm. The following diagram shows how the class loader implements the delegation hierarchy algorithm to load all the required classes:

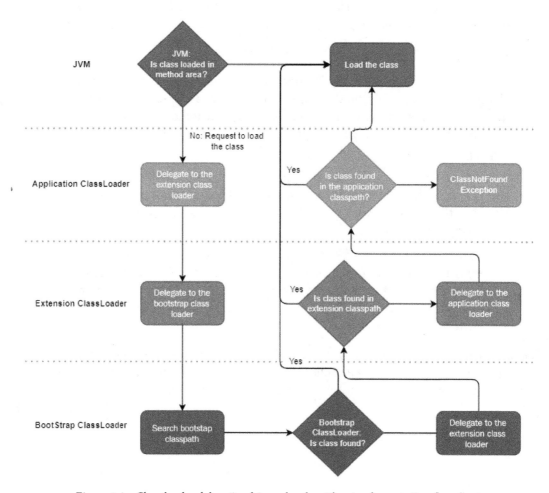

Figure 1.4 – Class loader delegation hierarchy algorithm implementation flowchart

Let's understand how this algorithm works:

1. JVM looks for the class in the method area (this will be discussed in detail later in this section). If it does not find the class, it will ask the application class loader to load the class into memory.

2. The application class loader delegates the call to the extension class loader, which in turn delegates to the bootstrap class loader.

3. The bootstrap class loader looks for the class in the bootstrap CLASSPATH. If it finds the class, it will load to the memory. If it does not find the class, control is delegated to the extension class loader.

4. The extension class loader will try to find the class in the extension CLASSPATH. If it finds the class, it will load to the memory. If it does not find the class, control is delegated to the application class loader.

5. The application class loader will try to look for the class in CLASSPATH. If it does not find it, it will raise ClassNotFoundException, otherwise, the class is loaded into the method area, and the JVM will start using it.

Linking

Once the classes are loaded into the memory (into the method area, discussed further in the *Memory subsystem* section), the class loader subsystem will perform linking. The linking process consists of the following steps:

- **Verification**: The loaded classes are verified for their adherence to the semantics of the language. The binary representation of the class that is loaded is parsed into the internal data structure, to ensure that the method runs properly. This might require the class loader to load recursively the hierarchy of inherited classes all the way to java.lang.Object. The verification phase validates and ensures that the methods run without any issues.

- **Preparation**: Once all the classes are loaded and verified, JVM allocates memory for class variables (static variables). This also includes calling static initializations (static blocks).

- **Resolution**: JVM then resolves by locating the classes, interfaces, fields, and methods referenced in the symbol table. The JVM might resolve the symbol during initial verification (static resolution) or may resolve when the class is being verified (lazy resolution).

The class loader subsystem raises various exceptions, including the following:

- `ClassNotFoundException`
- `NoClassDefFoundError`
- `ClassCastException`
- `UnsatisfiedLinkError`
- `ClassCircularityError`
- `ClassFormatError`
- `ExceptionInInitializerError`

You can refer to the Java specifications for more details: `https://docs.oracle.com/en/java/javase`.

Initializing

Once all the classes are loaded and symbols are resolved, the initialization phase starts. During this phase, the classes are initialized (new). This includes initializing the static variables, executing static blocks, and invoking reflective methods (`java.lang.reflect`). This might also result in loading those classes.

Class loaders load all the classes into the memory before the application can run. Most of the time, the class loader has to load the full hierarchy of classes and dependent classes (though there is lazy resolution) to validate the schematics. This is time-consuming and also takes up a lot of memory footprint. It's even slower if the application uses reflection and the reflected classes need to be loaded.

After learning about the class loader subsystem, let's now understand how the memory subsystem works.

Memory subsystem

The memory subsystem is one of the most critical subsystems of the JVM. The memory subsystem, as the name suggests, is responsible for managing the allocated memory of method variables, heaps, stacks, and registers. The following diagram shows the architecture of the memory subsystem:

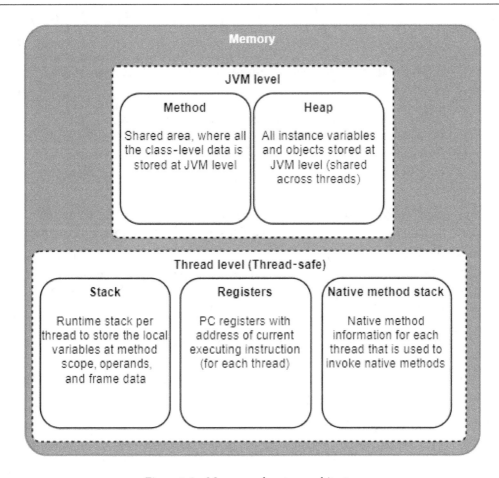

Figure 1.5 – Memory subsystem architecture

The memory subsystem has two areas: JVM level and thread level. Let's discuss each in detail.

JVM level

JVM-level memory, as the name suggests, is where the objects are stored at the JVM level. This is not thread-safe, as multiple threads might be accessing these objects. This explains why programmers are recommended to code thread-safe (synchronization) when they update the objects in this area. There are two areas of JVM-level memory:

- **Method**: The method area is where all the class-level data is stored. This includes the class names, hierarchy, methods, variables, and static variables.

- **Heap**: The heap is where all the objects and the instance variables are stored.

Thread level

Thread-level memory is where all the thread-local objects are stored. This is accessible/visible to the respective threads, hence it is thread-safe. There are three areas of the thread-level memory:

- **Stack**: For each method call, a stack frame is created, which stores all the method-level data. The stack frame consists of all the variables/objects that are created within the method scope, operand stack (used to perform intermediate operations), the frame data (which stores all the symbols corresponding to the method), and exception catch block information.

- **Registers**: PC registers keep track of the instruction execution and point to the current instruction that is being executed. This is maintained for each thread that is executing.

- **Native Method Stack**: The native method stack is a special type of stack that stores the native method information, which is useful when calling and executing the native methods.

Now that the classes are loaded into the memory, let's look at how the JVM execution engine works.

JVM execution engine subsystem

The JVM execution engine is the core of the JVM, where all the execution happens. This is where the bytecodes are interpreted and executed. The JVM execution engine uses the memory subsystem to store and retrieve the objects. There are three key components of the JVM execution engine, as shown:

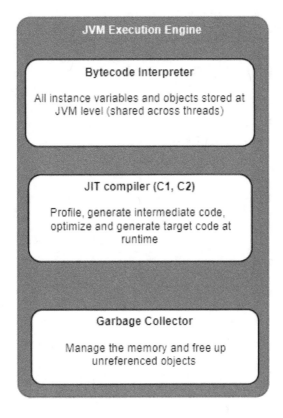

Figure 1.6 – JVM execution engine architecture

We will talk about each component in detail in the following sections.

Bytecode interpreter

As mentioned earlier in this chapter, bytecode (.class) is the input to the JVM. The JVM bytecode interpreter picks each instruction from the .class file and converts it to machine code and executes it. The obvious disadvantage of interpreters is that they are not optimized. The instructions are executed in sequence, and even if the same method is called several times, it goes through each instruction, interprets it, and then executes.

JIT compiler

The JIT compiler saves the day by profiling the code that is being executed by interpreters, identifies areas where the code can be optimized and compiles them to target machine code, so that they can be executed faster. A combination of bytecode and compiled code snippets provide the optimum way to execute the class files.

The following diagram illustrates the detailed workings of JVM, along with the various types of JIT compilers that the JVM uses to optimize the code:

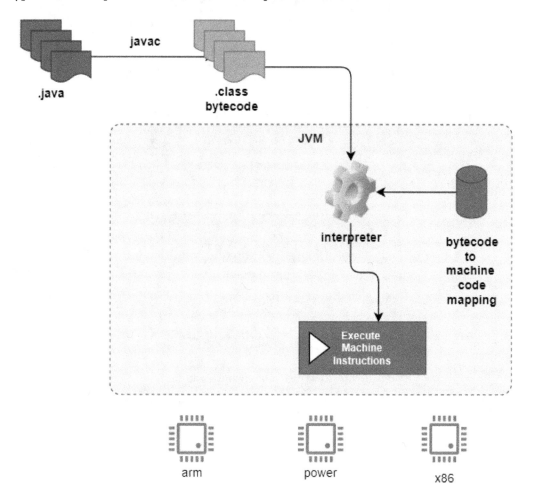

Figure 1.7 – The detailed working of JVM with JIT compilers

Let's understand the workings shown in the previous diagram:

1. The JVM interpreter steps through each bytecode and interprets it with machine code, using the bytecode to machine code mapping.

2. JVM profiles the code consistently using a counter, to count the number of times a code is executed, and if the counter reaches a threshold, it uses the JIT compiler to compile that code for optimization and stores it in the code cache.

3. JVM then checks whether that compilation unit (block) is already compiled. If JVM finds a compiled code in the code cache, it will use the compiled code for faster execution.

4. JVM uses two types of compilers, the C1 compiler and the C2 compiler, to compile the code.

As illustrated in *Figure 1.7*, the JIT compiler brings in optimizations by profiling the code that is running and, over a period of time, it identifies the code that can be compiled. The JVM runs the compiled snippets of code instead of interpreting the code. It is a hybrid method of running interpreted code and compiled code.

JVM introduced two types of compilers, C1 (client) and C2 (server), and the recent versions of JVM use the best of both for optimizing and compiling the code at runtime. Let's understand these types better:

- **C1 compiler**: A performance counter was introduced, which counted the number of times a particular method/snippet of code is executed. Once a method/code snippet is used a particular number of times (threshold), then that particular code snippet is compiled, optimized, and cached by the C1 compiler. The next time that code snippet is called, it directly executes the compiled machine instructions from the cache, rather than going through the interpreter. This brought in the first level of optimization.

- **C2 compiler**: While the code is getting executed, the JVM will perform runtime code profiling and come up with code paths and hotspots. It then runs the C2 compiler to further optimize the hot code paths. This is also known as a hotspot.

C1 is faster and good for short-running applications, while C2 is slower and heavy, but is ideal for long-running processes such as daemons and servers, so the code performs better over time.

In Java 6, there is a command-line option to use either C1 or C2 methods (with the command-line arguments -client (for C1) and -server (for C2)). In Java 7, there is a command-line option to use both. Since Java 8, both C1 and C2 compilers are used for optimization as the default behavior.

There are five tiers/levels of compilation. Compilation logs can be generated to understand which Java method is compiled using which compiler tier/level. The following are the five tiers/levels of compilation:

- Interpreted code (level 0)

- Simple C1 compiled code (level 1)

- Limited C1 compiled code (level 2)

- Full C1 compiled code (level 3)

- C2 compiled code (level 4)

Let's now look at the various types of code optimizations that the JVM applies during compilation.

Code optimizations

The JIT compiler generates the internal representation of the code that is being compiled to understand the semantics and syntax. These internal representations are tree data structures, on which the JIT will then run the code optimization (as multiple threads, which can be controlled with the XcompilationThreads options from the command line).

The following are some of the optimizations that the JIT compilers perform on the code:

- **Inlining**: One of the most common programming practices in object-oriented programming is to access the member variables through getter and setter methods. Inlining optimization replaces these getter/setter methods with actual variables. The JVM also profiles the code and identifies other small method calls that can be inlined to reduce the number of method calls. These are known as hot methods. A decision is taken based on the number of times that the method is called and the size of the method. The size threshold used by JVM to decide inlining can be modified using the -XX:MaxFreqInlineSize flag (by default, it is 325 bytes).

- **Escape analysis**: The JVM profiles the variables to analyze the scope of the usage of the variables. If the variables don't escape the local scope, it then performs local optimization. Lock Elision is one such optimization, where the JVM decided whether a synchronization lock is really required for the variable. Synchronization locks are very expensive to the processor. The JVM also decides to move the object from the heap to the stack. This has a positive impact on memory usage and garbage collection, as the objects are destroyed once the method is executed.

- **DeOptimization**: DeOptimization is another critical optimization technique. The JVM profiles the code after optimization and may decide to deoptimize the code. Deoptimizations will have a momentary impact on performance. The JIT compiler decides to deoptimize in two cases:

    a. **Not Entrant Code**: This is very prominent in inherited classes or interface implementations. JIT may have optimized, assuming a particular class in the hierarchy, but over time when it learns otherwise, it will deoptimize and profile for further optimization of more specific class implementations.

    b. **Zombie Code**: During Not Entrant code analysis, some of the objects get garbage collected, leading into code that may never be called. This code is marked as zombie code. This code is removed from the code cache.

Apart from this, the JIT compiler performs other optimizations, such as control flow optimization, which includes rearranging code paths to improve efficiency and native code generation to the target machine code for faster execution.

JIT compiler optimizations are performed over a period of time, and they are good for long-running processes. We will be going into a detailed explanation on JIT compilation in *Chapter 2, JIT, Hotspot, and GraalVM*.

Java ahead-of-time compilation

The ahead-of-time compilation option was introduced with Java 9 with `jaotc`, where a Java application code can be directly compiled to generate final machine code. The code is compiled to a target architecture, so it is not portable.

Java supports running both Java bytecode and AOT compiled code together in an x86 architecture. The following diagram illustrates how it works. This is the most optimum code that Java can generate:

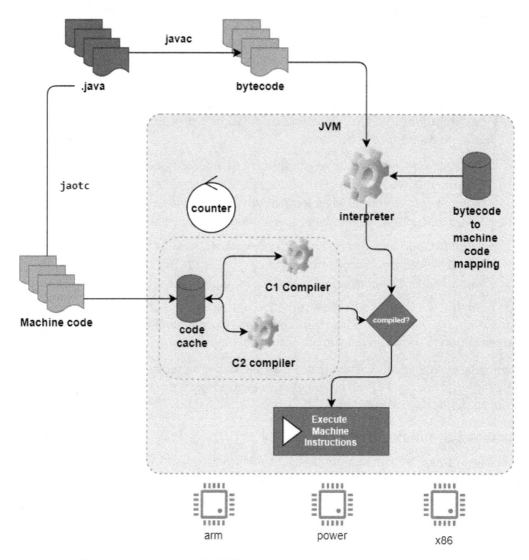

Figure 1.8 – The detailed workings of JVM JIT time compilers along with the ahead-of-time compiler

The bytecode will go through the approach that was explained previously (C1, C2). jaotc compiles the most used java code (like libraries) into machine code, ahead of time, and this is directly loaded into the code cache. This will reduce the load on JVM. The Java byte code goes through the usual interpreter, and uses the code from the code cache, if available. This reduces a lot of load on JVM to compile the code at runtime. Typically, the most frequently used libraries can be AOT compiled for faster responses.

Garbage collector

One of the sophistication of Java is its in-built memory management. In languages such as C/C++, the programmer is expected to allocate and de-allocate the memory. In Java, JVM takes care of cleaning up the unreferenced objects and reclaims the memory. The garbage collector is a daemon thread that performs the cleanup either automatically or can also be invoked by the programmer (System.gc() and Runtime.getRuntime().gc()).

Native subsystem

Java allows programmers to access native libraries. Native libraries are typically those libraries that are built (using languages such as C/C++) and used for a specific target architecture. **Java Native Interface (JNI)** provides an abstraction layer and interface specification for implementing the bridge to access the native libraries. Each JVM implements JNI for the specific target system. Programmers can also use JNI to call the native methods. The following diagram illustrates the components of the native subsystem:

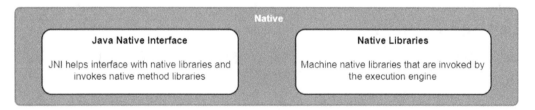

Figure 1.9 – Native subsystem architecture

The native subsystem provides the implementation to access and manage the native libraries.

JVM has evolved and has one of the most sophisticated implementations of a language VM runtime.

Summary

In this chapter, we started by learning what GraalVM is, followed by understanding how JVM works and its architecture, along with its various subsystems and components. Later on, we also learned how JVM combines the best of interpreters and the compiler approach to run Java code on various target architectures, along with how a code is compiled just-in-time with C1 and C2 compilers. Lastly, we learned about various types of code optimizations that the JVM performs.

This chapter provided a good understanding of the architecture of JVM, which will help us understand how the GraalVM architecture works and how it is built on top of JVM.

The next chapter will cover the details of how JIT compilers work and help you understand how Graal JIT builds on top of JVM JIT.

Questions

1. Why is Java code interpreted to bytecode and later compiled at runtime?

2. How does JVM load the appropriate class files and link them?

3. What are the various types of memory areas in JVM?

4. What is the difference between the C1 compiler and the C2 compiler?

5. What is a code cache in JVM?

6. What are the various types of code optimizations that are performed just in time?

Further reading

- *Introduction to JVM Languages*, by Vincent van der Leun, Packt Publishing (`https://www.packtpub.com/product/introduction-to-jvm-languages/9781787127944`)

- *Java Documentation and Specification*, by Oracle (`https://docs.oracle.com/en/java/`)

2
JIT, HotSpot, and GraalJIT

In the previous chapter, we learned about C1 and C2 compilers and the kind of code optimizations and de-optimizations that C2 compilers perform at runtime.

In this chapter, we will deep dive into the C2 just-in-time compilation and introduce Graal's just-in-time compilation. **Just-In-Time (JIT)** compilation is one of the key innovations that enabled Java to compete with traditional **ahead-of-time (AOT)** compilers. As we learned in the previous chapter, JIT compilation evolved with C2 compilers in JVM. The C2 JIT compiler constantly profiles code execution and applies various optimizations and de-optimizations at runtime to compile/recompile the code.

This chapter will be a hands-on session, where we will take a sample code and analyze how the C2 JIT compiler works and introduce Graal JIT.

In this chapter, we will cover the following topics:

- Understand how the JIT compiler works
- Learn how code is optimized by JIT by identifying HotSpots
- Use profiling tools to demonstrate how the JIT compiler works
- Understand how GraalVM JIT works on top of JVM JIT

By the end of this chapter, you will have a clear understanding of the internal workings of the JIT compiler and how GraalVM extends it further. We will be using sample Java code and profiling tools such as JITWatch to gain a deeper understanding of how JIT works.

Technical requirements

To follow the instructions given in this chapter, you will require the following:

- All the source code referred to in this chapter can be downloaded from `https://github.com/PacktPublishing/Supercharge-Your-Applications-with-GraalVM/tree/main/Chapter02`.

- Git (`https://github.com/git-guides/install-git`)

- Maven (`https://maven.apache.org/install.html`)

- OpenSDK (`https://openjdk.java.net/`) and JavaFX (`https://openjfx.io/`)

- JITWatch (`https://www.jrebel.com/blog/understanding-java-jit-with-jitwatch#:~:text=JITWatch%20is%20a%20log%20analyser,to%20the%20Adopt%20OpenJDK%20initiative`)

- The Code in Action video for this chapter can be found at `https://bit.ly/3w7uWlu`.

Setup environment

In this section, we will set up all the prerequisite tools and environments that are required to follow on with the rest of the chapter.

Installing OpenJDK Java

You can install OpenJDK from `https://openjdk.java.net/install/`. This URL has detailed instructions on how to install OpenJDK. We also require JavaFX. Please refer to `https://openjfx.io/` for more details on how to install JavaFX.

Installing JITWatch

JITWatch is one of the most widely used log analysis and visualization tools for understanding the behavior of the JIT compiler. This is also widely used in analyzing the code and identifying opportunities for better performance tuning.

JITWatch is an active open source project hosted at `https://github.com/AdoptOpenJDK/jitwatch`.

The typical commands for installing JITWatch are as follows:

```
git clone git@github.com:AdoptOpenJDK/jitwatch.git
cd jitwatch
mvn clean install -DskipTests=true
./launchUI.sh
```

Taking a deep dive into HotSpot and the C2 JIT compiler

In the previous chapter, we walked through the evolution of JVM and how the C2 JIT compiler evolved. In this section, we will dig deeper into the JVM C2 JIT compiler. Using sample code, we will go through the optimizations that the JIT compiler performs at runtime. To appreciate the Graal JIT compiler, it is very important to understand how the C2 JIT compiler works.

Profile-guided optimization is the key principle for JIT compilers. While AOT compilers can optimize the static code, most of the time, that is just not good enough. It's important to understand the runtime characteristics of the application to identify opportunities for optimization. JVM has a built-in profiler that dynamically instruments the application to profile some key parameters and to identify opportunities for optimizations. Once identified, it will compile that code to the native language and switch from running the interpreted code to faster-compiled code. The optimizations are based on profiling and educated assumptions that are made by JVM. If any of these assumptions are incorrect, JVM will de-optimize and switch back to running interpreted code. This is called **Mixed Mode Execution**.

The following diagram shows a flow of how JVM performs the profile-guided optimizations and switches between different modes of execution:

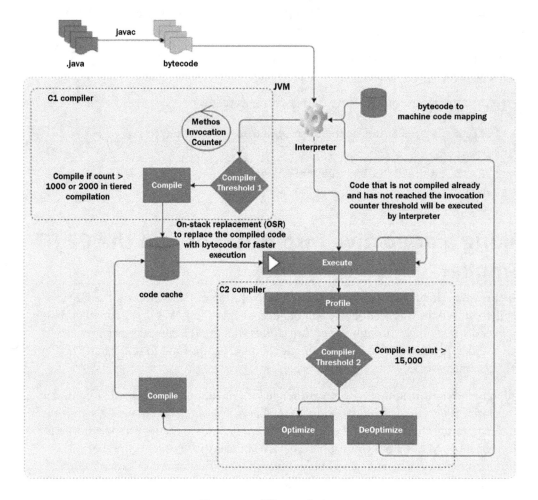

Figure 2.1 – JIT compilation

The Java source code (`.java`) is compiled to bytecode (`.class`), which is an intermediate representation of the code. JVM starts to run the bytecode with the inbuilt interpreter. The interpreter uses a bytecode to machine code mapping, converts the bytecode instructions to machine code one statement at a time, and then executes it.

While JVM executes these instructions, it also tracks the number of times a method is invoked. When the method invocation count of a particular method exceeds the compiler threshold, it spins off a compiler to compile that method on a separate compilation thread. There are two types of compilers that JVM uses to compile the code: C1 (client) and C2 (server) JIT compilers. The compiled code is stored in the code cache, so that the next time that method is invoked, JVM will execute the code from the code cache instead of interpreting it. JIT compilers perform various optimizations to the code, and hence, over time, the application performs better. The rest of this section will walk through these various components in detail.

Code cache

The code cache is an area in JVM where JVM stores the compiled native methods (also referred to as nmethod). The code cache is set to a static size and it might become full after a while. Once the code cache is full, JVM cannot compile or store any more code. It is very important to tune the code cache for optimum performance. Four key parameters help us to fine-tune JVM performance, with the optimum code cache:

- -XX:InitialCodeCacheSize: The initial size of the code cache. The default size is 160 KB (varies based on the JVM version).

- -XX:ReservedCodeCacheSize: The maximum size the code cache can grow to. The default size is 32/48 MB. When the code cache reaches this limit, JVM will throw a warning: CodeCache is full. Compiler has been disabled.. JVM offers the UseCodeCacheFlushing option to flush the code cache when the code cache is full. The code cache is also flushed when the compiled code is not hot enough (when the counter is less than the compiler threshold).

- -XX:CodeCacheExpansionSize: This is the expansion size when it scales up. Its default value is 32/64 KB.

- -XX:+PrintCodeCache: This option can be used to monitor the usage of the code cache.

Since Java 9, JVM segments the code cache into three segments:

- **Non-native method segment**: This segment contains the JVM internal code (such as the bytecode interpreter). The default size of this segment is 5 MB. This can be changed using the `-XX:NonNMethodCodeHeapSize` flag.

- **Profiled code**: This segment contains the compiled profiled code. This code is not completely optimized and has instrumentation that the profiler uses to optimize the code further. The default size is 122 MB. This can be changed using the `-XX:ProfiledCodeHeapSize` flag.

- **Non-profiled code**: This is the fully optimized code, where even the instrumentation is removed. This can be changed using the `-XX:NonProfiledCodeHeapSize` flag.

Compiler threshold

The compilation threshold is a factor that helps JVM decide when to perform JIT compilations. When JVM detects that a method execution has reached a compilation threshold, JVM will instigate the appropriate compiler to perform the compilation (more on this later in this section, where we will walk through the various types of JIT compilers and tiered compilation).

Deciding the compilation threshold is based on two key variables. These variables come with a default value for each JVM, but can also be changed with appropriate command-line arguments. These variables are very critical in tuning the performance of JVM and should be used carefully. These two variables are as follows:

- **Method invocation counter**: This counts the number of times a particular method is invoked.

- **Loop counter**: This refers to the number of times a particular loop has completed execution (what is referred to as branching back). Sometimes, this is also referred to as Backedge Threshold or Backedge Counter.

JVM profiles these two variables at runtime and, on this basis, decides whether that method/loop needs to be compiled. When a compilation threshold is reached, JVM spins off a compilation thread to compile that particular method/loop.

The compilation threshold can be changed using the `-XX:CompilationThreshold=N` flag as an argument while executing the code. The default value of N is `1500` for the client compiler and `10000` for the server compiler.

On-stack replacement

The methods that reach the compilation threshold are compiled by the JIT compilers, and the next time the method is called, the compiled machine code is called. This improves performance over time. However, in cases of long-running loops that reach the loop counter threshold (Backedge Threshold), the compilation thread initiates code compilation. Once the code that is in the loop is compiled, the execution is stopped and resumed with the compiled code frame. This process is called **On-Stack Replacement (OSR)**. Let's look at the following example.

The following code snippet just discusses how OSR works. To keep it simple, the code simply shows a long-running loop where we are just calculating the total number of times the loop runs. In this case, the main() method is never entered, so even after the compilation threshold is reached and the code is compiled, the compiled code cannot be used as the interpreter continues to execute, unless the code is replaced. This is where OSR helps in optimizing such code:

```java
public class OSRTest {
    public static void main(String[] args) {

        int total = 0;
        //long running loop
        for(int i=0; i < 10000000; i++) {

            //Perform some function
            total++;
        }
        System.out.println("Total number of times is "+ total);

    }
}
```

The following flowchart shows how the OSR works in this case:

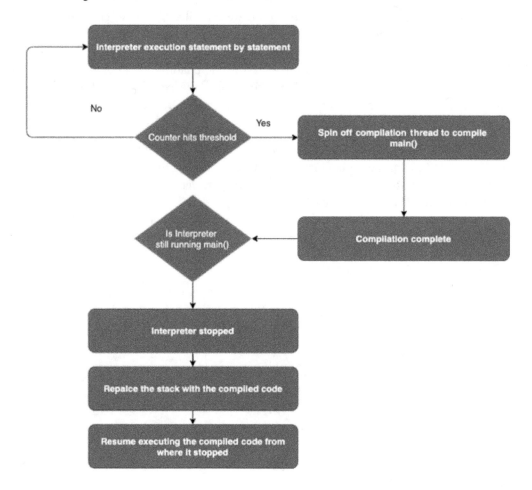

Figure 2.2 – OSR flowchart

Let's see how this works:

1. The interpreter starts executing the code.

2. When the compiler threshold is reached, JVM spins off a compiler thread to compile the method. In the meantime, the interpreter continues to execute the statement.

3. Once the compilation thread comes back with the compiled code (compilation frame), JVM checks whether the interpreter is still executing the code. If the interpreter is still executing the code, it will pause and perform OSR, and execution starts with the compiled code.

When we run this code with the -XX:PrintCompilation flag on, this is the output that shows that JVM performed OSR (the % attribute indicates that it performed OSR):

```
455 1114      4          java.lang.String::replace (42 bytes)
456 1115 %    3          OSRTest::main @ 4 (32 bytes)
```

Figure 2.3 – OSR log screenshot

Please refer to the next section to understand the log format in detail.

XX:+PrintCompilation

XX:+PrintCompilation is a very powerful argument that can be passed to understand how the JIT compilers are kicking in and optimizing the code. Before we run our code with this argument, let's first understand the output format.

XX:+PrintCompilation produces a log list of parameters separated by blank spaces in the following format:

```
<Timestamp> <CompilationID> <Flag> <Tier>
<ClassName::MethodName> <MethodSize> <DeOptimization Performed
if any>
```

Here is an example snapshot of the output:

```
615 1346 %    4          CountCharacters::main @ 51 (177 bytes)
637 1344 %    3          CountCharacters::main @ 51 (177 bytes)    made not entrant
761 1346 %    4          CountCharacters::main @ 51 (177 bytes)    made not entrant
```

Figure 2.4 – Print compilation log format

Let's look at what these parameters mean:

- Timestamp: This is the time in milliseconds since JVM started.

- CompilationID: This is an internal identification number used by JVM in the compilation queue. This will not necessarily be in a sequence, as there are background compilation threads that might reserve some of the IDs.

- Flags: The compiler flags are very important parameters that are logged. This suggests what attributes are applied by the compiler. JVM prints a comma-separated string of five possible characters to indicate five different attributes that are applied to the compiler. If none of the attributes are applied, it is shown as a blank string. The five attributes are as follows:

    a. **On-Stack Replacement**: This is represented by the % character. OSR is explained earlier in this section. This attribute suggests that an OSR compilation is triggered as the method is looping over a large loop.

    b. **Exception Handler**: This is represented by the ! character. This indicates that the method has an exception handler.

    c. **Synchronized method**: This is represented by the s character. This indicates that the method is synchronized.

    d. **Blocking Mode**: This is represented by the b character. This indicates that the compilation occurred in blocking mode. This means that the compilation did not happen in the background.

    e. **Native**: This is represented by the n character. This indicates that the code is compiled to the native method.

- Tier: This indicates which tier of compilation is performed. Refer to the *Tiered compilation* section for more details.

- MethodName: This column lists the method that is being compiled.

- MethodSize: This is the size of the method.

- Deoptimization performed: This shows any de-optimizations that may be performed. We will discuss this in detail in the next section.

Tiered compilation

In the previous chapter, we briefly covered compilation tiers/levels. In this section, we will go into more details. Client compilers kick in early when the compiler threshold is reached. The server compiler kicks in based on the profiling. The recent versions of JVM use a combination of both compilers for optimum performance. However, the user can specifically use one of the compilers with the -client, -server, or -d64 arguments. The default behavior of JVM is to use tiered compilation, which is the most optimum JIT compilation. With tiered compilation, the code is first compiled by the client compiler and, based on the profiling, if the code gets hotter (hence the name HotSpot), the server compiler kicks in and recompiles the code. This process was explained in the previous section by means of the flowchart.

Tiered compilation brings in more optimization as the code gets complicated and runs for longer. There are instances where JIT compilation runs more optimally and faster than AOT compilation. While AOT compilation brings in optimization, during the build phase, it does not have the intelligence to optimize/deoptimize itself based on runtime profiling. Runtime profiling, optimizing, and deoptimizing are the key advantages of JIT compilation.

There are three versions of JIT compilers:

- **C1**: The 32-bit client version is for applications that we are running on 32-bit operating systems. For a 64-bit operating system, both 32-bit and 64-bit versions of the JIT compilers can be used. Typically, 32-bit versions (both client and server) are ideal for smaller heap sizes (smaller footprint). This version of the compiler can be explicitly invoked using the -client argument:

```
java -client -XX:+PrintCompilation <Class File>
```

  -XX:PrintCompilation logs the compilation process to the console. This helps in understanding how the compiler is working.

- **C2 32-bit**: The 32-bit server version JIT compiler is ideal for 32-bit operating systems and applications that have a smaller footprint and do not perform extensive operations on long or double variables. This version of the compiler can be explicitly invoked using the -server argument.

- **C2 64-bit**: The 64-bit server version JIT compiler is for 64-bit operating systems and is ideal for large applications. They have a larger footprint and are not as fast as 32-bit compilers. However, 64-bit compilers can perform faster and better. This version of the compiler can be explicitly invoked using the -d64 argument.

Server compilers are up to 4x slower in compiling than client compilers. However, they do generate a faster running application (up to 2x).

There are five tiers/levels of compilation levels as listed next. A compilation log can be used to find which method was compiled to what level, by means of compilation print:

- **Level 0 – Interpreted code**: This the standard interpreter mode, where the JIT is still not activated. The JIT gets activated based on the compilation threshold.

- **Level 1 – Simple C1 compiled code**: This is a basic no-profile compilation of the code. The compiled code will not have any instrumentation.

- **Level 2 – Limited C1 compiled code**: In this level, basic counters are instrumented. The counter will help JVM decide to move to the next level, L2. Sometimes, when the C2 compiler is busy, JVM will use this level as an intermediate before promotion to Level 3.

- **Level 3 – Full C1 compiled code**: In this level, the code is fully instrumented and profiled. This detailed profiling will help decide further optimization with L4. This level adds up to 25-30% of overhead to the compiler and performance.

- **Level 4 – C2 compiled code**: This is the most optimized compilation of the code, where all the optimization is applied. However, while profiling, if JVM finds that the context of optimization has changed, it will deoptimize and replace the code with L0 or L1 (for trivial methods).

Let's now look at how the Java HotSpot compiler performs tiered compilation. The following diagram shows the various tiers and flow patterns of compilation:

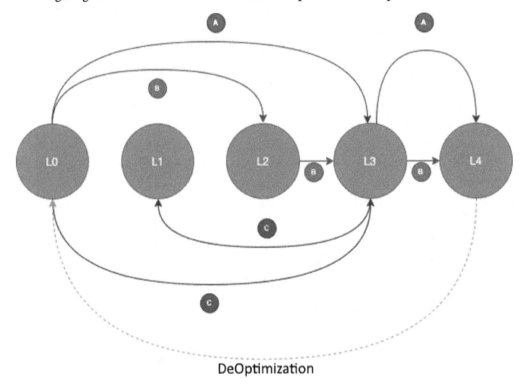

Figure 2.5 – Tiered compilation patterns

Let's understand what each flow indicates:

- **A**: This is the normal pattern of how JVM works. All the code starts with L0 and escalates to L3 when the compilation threshold is reached. At L3, the code is compiled with complete detailed profiling instrumentation. The code is then profiled at runtime, when it reaches the threshold, and then the code is re-compiled with the C2 compiler (L4), with maximum optimization. C2 compilers require detailed data regarding the control flow so as to take decisions concerning optimization. Later in this section, we will walk through all the optimizations that the C2 compiler (JIT) performs. It is possible, however, that the optimizations are invalid, due to changes in the flows or the context of optimization. In this case, JVM will deoptimize and bring it back to L0.

- **B: C2 Busy**: C2 compilation is performed on a separate compilation thread and the compilation activities are queued. When the compilation threads are all busy, JVM will not follow the normal flow, as this may affect the overall performance of the application. Instead, JVM will escalate to L2, where at least the counters are instrumented, and at a later point, when the code reaches the higher threshold, it will escalate to L3 and L4. At any point, JVM can deoptimize or invalidate the compiled code.

- **C: Trivial Code**: Sometimes, JVM will compile the code to L3 and realize that the code does not require any optimization, as it is very straightforward/simple, based on the profiling. In this case, it will bring it down to L1. That way, the execution of the code will be faster. The more we instrument the code, the more overhead we are putting on the execution. It is typically observed that L3 adds anywhere between 20-30% overhead to execution, due to instrumentation code.

We can look at how JVM behaves using the `-XX:+PrintCompilation` option. Here is an example of a normal flow:

```
public class Sample {
    public static void main(String[] args) {
        Sample samp = new Sample();
        while (true) {
        for(int i=0; i<1000000; i++) {
            samp.performOperation();
        }
    }
}
}
```

```java
public void performOperation() {
    int sum = 0;
    int x = 100;
    performAnotherOperation();

}

public void performAnotherOperation() {
    int a = 100;
    int b = 200;
    for(int i=0; i<1000000; i++) {
        int x = a + b;
        int y = (24*25) + x;
        int z = (24*25) + x;
    }
}
}
```

For this code, when we execute java with -XX:+PrintCompilation, the following
log is generated on the console. The log can be redirected to a log file using the
+LogCompilation flag:

```
149   169        3        java.lang.String::indexOf (7 bytes)
149   165        3        jdk.internal.misc.Unsafe::putObjectRelease (9 bytes)
149   167        3        java.util.concurrent.ConcurrentHashMap::setTabAt (20 bytes)
154   159 %      3        Sample::performAnotherOperation @ 9 (43 bytes)   made not entrant
154   163        4        java.lang.String::charAt (25 bytes)
157    22        3        java.lang.String::charAt (25 bytes)    made not entrant
157   171        4        Sample::performAnotherOperation (43 bytes)
158   160        3        Sample::performAnotherOperation (43 bytes)   made not entrant
158   170        4        java.lang.String::isLatin1 (19 bytes)
158   172        3        Sample::performOperation (10 bytes)
158     7        3        java.lang.String::isLatin1 (19 bytes)   made not entrant
158   173        4        Sample::performOperation (10 bytes)
160   172        3        Sample::performOperation (10 bytes)   made not entrant
160   174 %      3        Sample::main @ 10 (29 bytes)
160   175        3        Sample::main (29 bytes)
161   176 %      4        Sample::main @ 10 (29 bytes)
165   174 %      3        Sample::main @ 10 (29 bytes)   made not entrant
165   176 %      4        Sample::main @ 10 (29 bytes)   made not entrant
165   177 %      3        Sample::main @ 10 (29 bytes)
166   178 %      4        Sample::main @ 10 (29 bytes)
171   177 %      3        Sample::main @ 10 (29 bytes)   made not entrant
```

Figure 2.6 – Log showing tiered compilation

In this screenshot, you can see how the main() method moves from L0->L3->L4, which is the normal flow (A). As JVM performs optimizations and de-optimizations, jumping between these various levels of compilation, it reaches the most optimum, stable point. This is one of the greatest advantages that JIT compilers have over AOT compilers. The JIT compiler uses the runtime behavior to optimize the code execution (not just the semantic/static code optimizations). If you run this with JITWatch, we can see a clearer representation. The following screenshot shows the JITWatch tool compile chain when we run it by the Sample.java snippet:

Figure 2.7 – JITWatch tiered compilation

The previous screenshot shows that Sample::main() is compiled with the C1-L3 compiler. Sample::Sample() (default constructor) is inlined and Sample::performOperation() is also inlined into Sample::main(). Sample::performAnotherOperation() is also compiled. This is the first level of optimization:

```
JITWatch Tiered Compiliation for Sample::main() method
```

The following screenshot shows how various compilers are run on each of the methods:

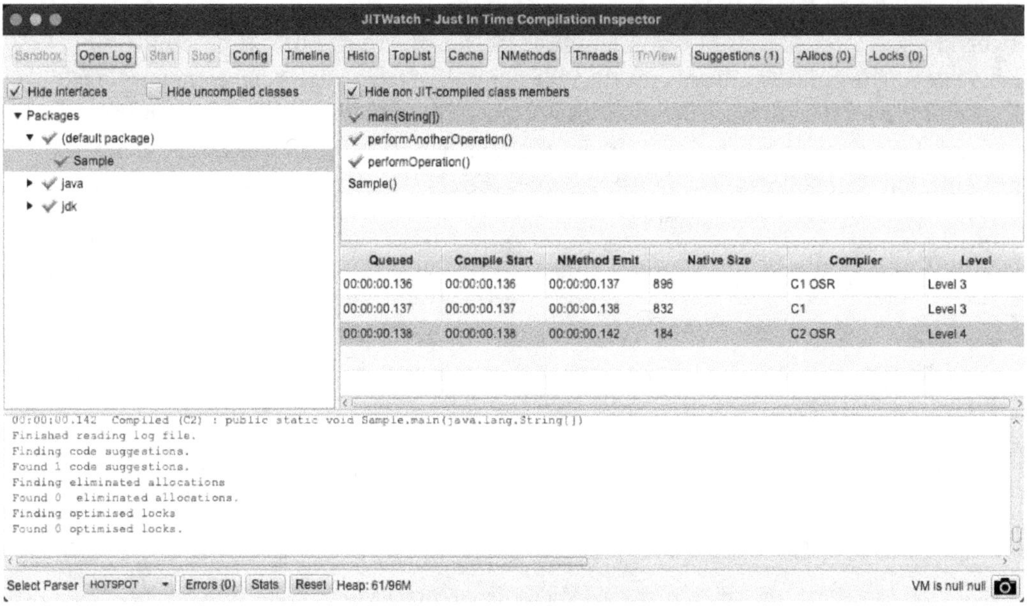

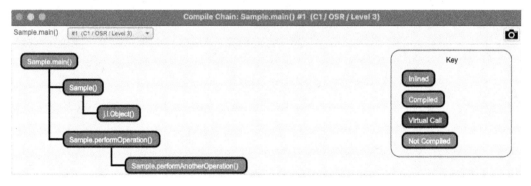

Figure 2.8 – JITWatch tiered compilation of main()

This screenshot shows how the `main()` method is optimized. Since the `main()` method has a long loop, OSR has occurred two times: once when the C1 compiled code is replaced, and the second time when the C2 compiled code is replaced. In each case, it has performed inlining. You can see what optimizations the C1 and C2 compilers performed in the following screenshot:

Figure 2.9 – JITWatch tiered compilation of main() – OSR-L3

In the previous screenshot, we can see that `Sample::performAnotherOperation()` is compiled and `Sample::performOperation()` is inlined into `Sample::main()`. The following screenshot shows further optimization that is performed by inlining `Sample:performAnotherOperation()` into `Sample::performOperation()`:

Figure 2.10 – JITWatch tiered compilation of main() – OSR-L4

Let's now look at how the JIT compiler optimizes the `Sample::performAnotherOperation()` method:

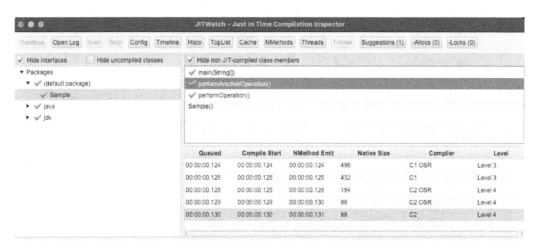

Figure 2.11 – JITWatch tiered compilation of performAnotherOperation()

As we can see in the previous screenshot, `Sample::performAnotherOperation()` has gone through various optimizations and OSRs, as it runs a long loop. The code is inlined into `Sample::performOperation()` as it hits the compiler threshold. The following screenshots reveal how `Sample::performAnotherOperation()` is compiled and inlined into `Sample::performOperation()`.

Let's now look at how the JIT compiler compiles the `Sample::performOperation()` method:

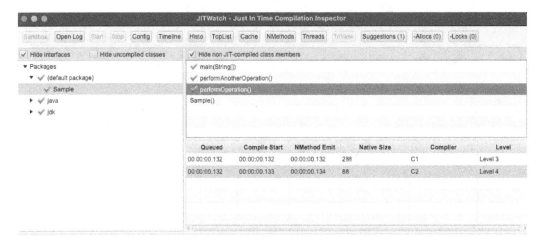

Figure 2.12 – JITWatch tiered compilation of performOperation()

The following screenshot shows the C1 compilation chain view for the `performOperation()` method:

Figure 2.13 – JITWatch tiered compilation of performOperation() – C1 compilation chain view

The previous screenshot shows that `Sample::performAnotherOperation()` is compiled as it hits the compiler threshold, and the following screenshot shows that the compiled code is inlined into `Sample::performOperation()`:

Figure 2.14 – JITWatch tiered compilation of performOperation() – C2 compilation chain view

JITWatch can be used to gain detailed understanding of how the C1 an C2 compilers behave, and how the optimizations are performed. This helps in reflecting on the application code, and proactively updating the source code for better runtime performance. To get a better understanding of how the C2 compiler optimizes the code, let's now look at the various types of code optimizations that JVM applies during compilation.

Understanding the optimizations performed by JIT

This section will cover the various optimization techniques that the JIT compilers employ at the various levels of compilation.

Inlining

Calling a method is an expensive operation for JVM. When the program calls a method, JVM has to create a new stack frame for that method, copy all the values into the stack frame, and execute the code. Once the method completes, the stack frame has to be managed post-execution. One of the best practices in object-oriented programming is to access the object members through access methods (getters and setters).

Inlining is one of the most effective optimizations performed by JVM. JVM replaces the method call with the actual content of the code.

If we run our previous code with the following command, we can see how JVM performs inlining:

```
java -XX:+PrintCompilation -XX:+UnlockDiagnosticVMOptions
-XX:+PrintInlining Sample
```

In this case, the call to the performOperation() method is replaced with the content of the inline main() method. After inlining effectively, the main() method will look like this:

```
public static void main(String[] args) {
        Sample samp = new Sample();
        while (true) {
        for(int i=0; i<1000000; i++) {
            samp.performOperation();
        }
    }
}
```

Inlining can be disabled using the -XX:-Inline flag.

JVM decides to inline the code, based on the number of calls made to the method and the size of the method, and if the method is frequently called (hot), and the size of the method is <325 bytes. Methods that are <35 bytes are inlined by default. These numbers can be changed with the -XX:+MaxFreqInlineSize and -XX:+MaxInlineSize flags from the command line, respectively.

Monomorphic, bimorphic, and megamorphic dispatch

Polymorphism is one of the key object-oriented concepts that provides a way to dynamically load classes based on a context, and the behavior is decided dynamically. Interfaces and inheritance are two of the most widely used polymorphism implementations. However, this comes with a performance overhead, as JVM loads the class/interface implementations dynamically. Inlining the implementations becomes a challenge.

One of the things that JVM profiles is the number of times a particular implementation is called and how many derived class/interface implementations really exist for a given base class or interface. If the profiler identifies only one implementation, then it's called monomorphic. If it finds two, then it's called bimorphic, and megamorphic means there are multiple implementations.

Based on the profiling, the JIT compiler identifies which specific derived class object (or interface implementation) is used and takes the decision on inlining that specific implementation, to overcome the performance overheads of polymorphism. Monomorphic and bimorphic are easy to inline. The JIT profiler tracks the execution paths and identifies in which context which implementation is used and performs inlining. Megamorphic implementations are complex to inline. The following code snippet shows polymorphism. We will use this code to understand the performance overhead:

```java
public interface Shape {
    String whichShapeAreYou();
}

public class Circle implements Shape {
    public String whichShapeAreYou() { return "I am Circle";}
}
public class Square implements Shape {
    public String whichShapeAreYou() { return "I am Square";}
}
public class Triangle implements Shape {
    public String whichShapeAreYou() { return "I am Triangle";}
}
public static void main(String[] args) {
    //Some code and logic here
    switch (circleType) {
        case 0:
            shape = new Circle();
            break;
        case 1:
            shape = new Square();
            break;
        case 2:
            shape = new Triangle();
            break;
        default:
            System.out.println("Invalid shape");
```

```
            break;
        }
    }
}
```

In the previous code, we have defined an interface called `Shape`, and we have three implementations of the interface, namely, `Circle`, `Square`, and `Triangle`. And, we are using a switch to initialize the right class. There are two optimization scenarios here:

- If the JIT knows that a particular implementation is used, it optimizes the code and might perform an inline. This is called a monomorphic dispatch.

- If, let's say, the decision is to be taken based on a particular variable or a configuration, JIT will profile, which is the most optimistic assumption it can take, and only those classes and inline them, and may also use an uncommon trap. In case the implementation class that is used is different from what is assumed, the JIT will deoptimize.

Dead code elimination

The JIT compiler identifies the code that is never executed or not required while profiling. This is called dead code, and the JIT compiler eliminates this from the execution. Modern IDEs identify dead code; this is purely based on the static code analysis performed. JIT compilers not only eliminate such trivial code, but also eliminate the code based on the control flow at runtime. Dead code elimination is one of the most effective ways to improve performance.

Let's take the following code as an example:

```
/**
 * DeadCodeElimination
 */
public class DeadCodeElimination {
    public void calculateSomething() {

        int[] arrayOfValues = new int[1000000];

        int finalTotalValue = 0;

        for (int i=0; i< arrayOfValues.length; i++) {
```

```
            finalTotalValue = calculateValue(arrayOfValues[i]);
        }

    //"Do some more activity here, but never use final Total
            count");
    }

    public int calculateValue(int value) {
        //use some formula to calucalte the value
        return value * value;
    }

    public static void main(String[] args) {
        DeadCodeElimination obj = new DeadCodeElimination();
        for (int i=0; i< 10000; i++) {
            obj.calculateSomething();
        }
    }

}
```

In this code, the `calculateSomething()` method has some logic. Let's look at the previous code snippet. The `finalTotalValue` variable is initialized and later, the total is calculated by calling the `calculateValue()` method in a loop, but assuming that `finalTotalValue` is never used after calculation. The initialization code, the array heap allocation code, and the loop that calls the `calculateValue()` method, are all dead code. JIT understands this at runtime and removes it completely.

JIT takes these decisions based on profiling and whether the code is reachable. It might remove unnecessary `if` statements (especially null checks; if the object is never seen to be null – this technique is sometimes referred to as Null Check Elimination). It will replace this with what is called "uncommon trap" code. If this execution ever reaches this trap code, it will then deoptimize.

Another case where "uncommon trap" code is placed by eliminating the code is by predicting branches. Based on the profiling, JIT assumes and predicts a branch code (if, switch, and so on) that may never be executed, and eliminates that code.

Common Subexpression Elimination is another effective technique that JIT uses to eliminate code. In this technique, an intermediate subexpression is removed to save the number of instructions.

Later, in the *Escape analysis* section, we will also see some code elimination techniques, based on the escape analysis that JIT performs.

Loop optimization – Loop unrolling

Loop unrolling is another effective optimization technique. This is more effective in smaller loop body and large number of iterations. The technique involves looking to reduce the iterations in the loop by replacing the code. Here is a very simple example:

```
for (int i=0; i< arrayOfValues.size; i++) {
    somefunction(arrayOfValues[i]);
}
```

This can be rolled into the following:

```
for (int i=0; i< arrayOfValues.size; i+=4) {
    somefunction (arrayOfValues[i]);
    somefunction (arrayOfValues[i+1]);
    somefunction (arrayOfValues[i+2]);
    somefunction (arrayOfValues[i+3]);
}
```

In this example, the JIT compiler decides to reduce the iterations by 1/4, by calling somefunction() four times instead of once. This has a significant performance improvement, as the number of jump statements goes down by 1/4. Of course, the decision on four is taken based on the size of the array so that the array reference does not go out of bounds.

Escape analysis

Escape analysis is one of the most advanced optimizations that the JIT compiler performs. This is controlled with the -XX:+DoEscapeAnalysis flag from the command line. This is enabled by default.

In the previous chapter, we went through the various memory areas in the *Memory subsystem* section. Heap and stack are two of the most important memory areas. The heap memory area is accessible across various threads in JVM. A heap is not thread-safe. When multiple threads access the data that is stored in the heap, it is recommended to write thread-safe code by obtaining synchronization locks. This blocks the other threads from accessing the same data. This has performance implications.

Stack memory is thread-safe, as it is allocated for that particular method call. Only the method thread has access to this area, hence there is no need to worry about obtaining synchronization locks or the absence of blocking threads.

The JIT compiler performs a detailed analysis of the code to identify code where we are allocating variables in the heap, but only using them in a specific method thread, and takes decisions on allocating these variables to the "stack area" instead of the "heap area." This is one of the most complex optimizations that JIT compilers perform and has a huge impact on performance. JIT might decide to store the variables in PC registers for even faster access.

JIT also looks for the use of synchronized and tracks. If it's called by a single thread, JIT decides to ignore synchronized. This has a significant impact on performance. StringBuffer is one of the objects that is thread-safe and has a lot of synchronized methods. If an instance of StringBuffer is not used outside a single method, JIT decides to ignore synchronized. This technique is sometimes referred to as "lock elision."

In cases where a synchronized lock cannot be ignored, the JIT compiler looks to combine the synchronized blocks. This technique is known as lock coarsening. This technique looks for subsequent synchronized blocks. Here is an example:

```
public class LockCoarsening {
    public static void main(String[] args) {
        synchronized (Class1.class) {
            ....
        }
        synchronized (Class1.class) {
            ....
        }
        synchronized (Class2.class) {
            ....
        }
    }
}
```

In this example, two subsequent synchronized blocks are trying to obtain a lock on the same class. The JIT compiler will combine these two blocks into one.

JIT performs a similar analysis for variables that are created with a loop and never used outside the loop. There is a very sophisticated technique called "scalar replacement," where the JIT profiles for objects that are created, but only a few member variables are used in the object that is not used. JIT will decide to stop creating the objects and replace them with the member variables directly. Here is a very simple example:

```
class StateStoring {
    final int state_variable_1;
    final int state_variable_2;
    public StateStoring(int val1, int val2) {
        this.state_variable_1 = val1;
        this.state_variable_2 = val2;
    }
}
```

The StateStoring class is a simple class, where we are storing the state of an object with two members – state_variable_1 and state_variable_2. JIT profiles this for various iterations and checks whether this object was created and never used outside the scope. It might decide not to even create the object, instead replacing the object getters and setters with actual scalars (local variables). That way, entire object creation and destruction (which is a very expensive process) can be avoided.

Here is a more advanced example, and this time let's see how JITWatch shows the escape analysis:

```
public class EscapeAnalysis2 {

    public void createNumberofObjects
    (int numberOfArraysToCreate, int numberOfCellsInArray) {
        for (int i=0; i< numberOfArraysToCreate; i++) {
            allocateObjects(numberOfCellsInArray);
        }
    }
```

```
    private void allocateObjects(int numberOfCellsInArray) {

        int[] arrayObj = new int[numberOfCellsInArray];

        for (int i=0; i< numberOfCellsInArray; i++) {
        //Heap allocation, which could have been easily a local
            stack allocation
            Integer dummyInt = new Integer(i);
            arrayObj[i] = dummyInt.intValue();
        }

        return;

    }

    public static void main(String[] args) {
        EscapeAnalysis2 obj = new EscapeAnalysis2();
        obj.createNumberofObjects(100000, 10);
    }

}
```

In this code snippet, the `allocateObjects()` method is creating an array (on the heap) and adding that value to the array. The `dummyInt` variable's scope is limited to the `for` loop in the `allocateObjects()` method. There is no need to have these objects created in the heap. After performing the escape analysis, JIT identifies that these variables can be put in a stack frame instead.

The following JITWatch screenshot demonstrates this:

Figure 2.15 – JITWatch escape analysis – 1

In this screenshot, the bytecode that allocates dummyInt is struck off to indicate that heap allocation for that variable is not required:

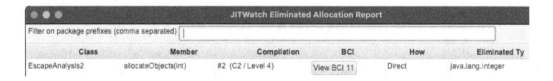

Figure 2.16 – JITWatch escape analysis – 2

The previous screenshot shows the optimization that is performed by C2/Level 4, where it removes allocation of the variable.

Deoptimization

In the previous section, we looked at various optimization techniques that the JIT compiler performs. The JIT compiler optimizes the code with some assumptions that it makes, based on the profiling. Sometimes, these assumptions may be not correct in a different context. When JIT stumbles upon these scenarios, it deoptimizes the code and goes back to using an interpreter to run the code. This is called Deoptimization and has an impact on performance.

There are two scenarios where Deoptimization occurs:

- When the code is "non-entrant"
- When the code is "zombie"

Let's understand these scenarios with the help of examples.

Non-entrant code

There are two cases where the code becomes non-entrant:

- **Assumptions made during polymorphism**: As we discussed in the section on monomorphic dispatch, polymorphism has a significant performance overhead on JVM. One of the optimizations that JIT performs is assuming a particular implementation of the interface/base class, and inlines that particular implementation of the interface/class. This is done based on the context and the control path that the JIT observed during profiling. When the assumption is invalid, JIT generates a deoptimization trap, and this optimized code is rendered "non-entrant". We can see when the JIT is making the code non-entrant with the -XX:+PrintCompilation flag.

- **Escalations that happen during tiered compilation**: When JIT decides to move between the level discussed in the *Tiered compilation* section, it marks the code that was optimized in the previous level as non-entrant, and the next level of the compilation process will optimize the code and replace the previous optimizations. This happens when JVM replaces the code that is compiled by C1 with the code that C2 has compiled. The following screenshot demonstrates an example when we run out Sample.java:

```
766  172      4      Sample::performOperation (10 bytes)
768  171      3      Sample::performOperation (10 bytes)    made not entrant
771  173 %    3      Sample::main @ 10 (27 bytes)
772  174      3      Sample::main (27 bytes)
773  175 %    4      Sample::main @ 10 (27 bytes)
778  173 %    3      Sample::main @ 10 (27 bytes)    made not entrant
778  175 %    4      Sample::main @ 10 (27 bytes)    made not entrant
```

Figure 2.17 – Tiered compilation escalation

In the preceding screenshot, we can see the tiered compilation in action (the third column shows the tier number) and the optimization that is done.

Zombie code

In most cases, some objects are created in the heap of the code that is marked as non-entrant. Once the GC reclaims all these objects, then JVM will mark the methods of those classes as zombie code. JVM then removes this compiled zombie code from the code cache. As we discussed in the *Taking a deep dive into hotspot and the C2 JIT* section, it's very important to keep the code cache optimum, as this has a significant impact on performance.

As we saw in tiered compilation, deoptimization is performed when any of the assumptions that the Java JIT made is challenged by the control flow at runtime. In the next section, we will briefly introduce the Graal JIT compiler, and how it plugs into JVM.

Graal JIT and the JVM Compiler Interface (JVMCI)

In the previous sections, as we walked through the various features and advancements that JIT compilers underwent, it is very clear that C2 is very sophisticated. However, C2 compiler implementation has its downsides. C2 is implemented in the C/C++ language. While C/C++ is fast, it is not type-safe and it does not have garbage collection. Hence, the code becomes very complex. C2 implementation is very complex, as it has become more and more complex to change the code for new enhancements and bug fixes.

In the meantime, Java has matured to run as fast as C/C++ in many cases. Java is type-safe with garbage collection. Java is simpler and easier to manage than C/C++. The key advantages of Java are its exception handling capabilities, memory management, better IDE/profiling, and tooling support. The JIT compiler is nothing but a program that takes in a bytecode, `byte[]`, optimizes it, compiles it, and returns an array of machine code, `byte[]`. This can easily be implemented in Java. What we need is a JVM interface that can provide the protocol for implementing the custom compiler logic. This will help open up JVM for the new implementations of JIT compilers.

JDK enhancement proposal JEP243 (`https://openjdk.java.net/jeps/243`) is a proposal to provide a compiler interface that will enable writing a compiler in Java and extending JVM to use it dynamically.

JEP243 was added in Java 9. This is one of the most significant enhancements to JVM. JVMCI is an implementation of JEP243. JVMCI provides the required extensibility to write our own JIT compilers. JVMCI provides the API that is required to implement custom compilers and configure JVM to call these custom compiler implementations. The JVMCI API provides the following capabilities:

- Access to VM data structures, which is required to optimize the code
- Managing the compiled code following optimization and deoptimization
- Callbacks from JVM to execute the compilation at runtime

A JVMCI can be executed with the following command-line flags:

```
-XX:+UnlockExperimentalVMOptions
-XX:+EnableJVMCI
-XX:+UseJVMCICompiler
-Djvmci.Compiler=<name of compiler>
```

Graal is an implementation of JVMCI, which brings all the key features and optimizations that are required for a modern Java runtime. Graal is wholly implemented in Java. Graal is much more than just a JIT compiler. Here is a quick comparison between the Graal JIT and the JVM JIT (C2):

JVM JIT (C2)	Graal JIT
Implemented in C/C++ and has matured and hardened over time.	Implemented in Java, still in its infancy.
Production grade.	Production grade since 19.0.0.
Complex code and has become tougher to maintain. This has reached its end of life.	Highly modular and brings in all the sophisticated modern Java programming language features. Easy to enhance, maintain, and manage.
The focus is to optimize Java source code.	This focus of Graal is much more than just JIT compilation. Targeting the cloud-native programming requirements of Polyglot; a smaller footprint, faster build, and run. Graal does not only focus on JVM-based (Scala, Kotlin, and so on) languages, but also on dynamic languages (Ruby, Python, JavaScript, and suchlike) as well as native languages (C/C++).

The next chapter will go into more detail on Graal architecture, and *Chapter 4, Graal Just-In-Time Compiler*, will go deeper into how Graal JIT works, and how it builds on top of Java JIT and brings in more advanced optimizations and support for Polyglot.

Summary

In this chapter, we went into a lot of detail on how the JIT compiler works and discussed the tiered compilation patterns that JVM uses to optimize the code. We also walked through various optimization techniques with a number of sample code examples. This provided a good understanding of the internal workings of JVM.

JVMCI provides the extensibility to build custom JIT compilers on JVM. Graal JIT is an implementation of JVMCI.

This chapter provided the basis for understanding how JVM works, and how JIT compilation optimizes the code at runtime. This is key in understanding how the Graal JIT compiler works.

In the next chapter, we will understand how the Graal VM architecture is built on the JVM architecture, and how it extends it to support Polyglot.

Questions

1.   What is a code cache?
2.   What are the various flags that can be used to optimize a code cache?
3.   What is the compiler threshold?
4.   What is on-stack replacement?
5.   What is tiered compilation? What are the various patterns of tiered compilation?
6.   What is inlining?
7.   What is monomorphic dispatch?
8.   What is loop unrolling?
9.   What is escape analysis?
10. What is Deoptimization?
11. What is JVMCI?

Further reading

- *Introduction to JVM Languages* (`https://www.packtpub.com/product/introduction-to-jvm-languages/9781787127944`)

- Java SDK documentation (`https://docs.oracle.com`)

- GraalVM documentation (`https://docs.oracle.com/en/graalvm/enterprise/19/guide/overview/compiler.html`)

- JITWatch documentation (`https://github.com/AdoptOpenJDK/jitwatch`)

Section 2: Getting Up and Running with GraalVM – Architecture and Implementation

This section will introduce Graal, GraalVM, Truffle, and the various architectural components of Graal. It also covers how Graal builds on top of JVM to provide a more optimized virtual machine for multiple language implementations. There is also a hands-on session on how to use Graal. This section contains the following chapters:

- *Chapter 3, GraalVM Architecture*
- *Chapter 4, Graal Just-In-Time Compiler*
- *Chapter 5, Graal Ahead-of-Time Compiler and Native Image*

3
GraalVM Architecture

In *Chapter 1*, *Evolution of Java Virtual Machine*, we took a detailed look at the JVM architecture. In *Chapter 2*, *JIT, HotSpot, and GraalJIT*, we went into more detail on how JVM JIT compilers work. We also looked at how JVM has evolved into an optimum HotSpot VM, with C1 and C2 JIT compilers.

While the C2 compiler is very sophisticated, it has become a very complex piece of code. GraalVM provides a Graal compiler, which builds on top of all the best practices from the C2 compiler, but it is built entirely from the ground up in Java. Hence, Graal JIT is more object-oriented, and has modern and manageable code, with the support of all of the modern integrated development environments, tools, and utilities to monitor, tune, and manage the code. GraalVM is much more than just the Graal JIT compiler. GraalVM brings in a larger ecosystem of tools, runtimes, and APIs to support multiple languages (Polyglot) to run on VMs, leveraging the most mature and hardened JIT compilation provided by Graal.

In this chapter, we will focus on the GraalVM architecture and its various components to achieve the most advanced, fastest, polyglot runtime for the cloud. We will also explore the cloud-native architectural patterns, and how GraalVM is the best platform for the cloud.

Before we get into the details of the GraalVM architecture, we will begin by learning the requirements of a modern technical architecture. Later in the chapter, as we go through each of the GraalVM architectural components, we will address these requirements.

In this chapter, we will cover the following topics:

- Reviewing modern architectural requirements

- Learning what the GraalVM architecture is

- Reviewing the GraalVM editions

- Understanding the GraalVM architecture

- An overview of the GraalVM microservices architecture

- An overview of various microservices frameworks that can build code for GraalVM

- Understanding how GraalVM addresses various non-functional aspects

By the end of this chapter, you will have a very clear understanding of the GraalVM architecture and how various components come together to provide a comprehensive VM runtime for polyglot applications.

Reviewing modern architectural requirements

Before we dig deeper into the GraalVM architecture, let's first understand the shortcomings of JVM and why we need a new architecture and approach. The older versions of JVM were optimized for traditional architectures, which were built for long-running applications that run in a data center, providing high throughput and stability (for example, monolith web application servers and large client-side applications). Some microservices are long-running, and Graal JIT will also provide the optimum solution. As we move to cloud-native, the whole architecture paradigm has shifted to componentized, modularized, distributed, and asynchronous architecture tuned to run efficiently with high scalability and availability requirements.

Let's break this down into more specific requirements for the modern cloud-native architectures.

Smaller footprint

The applications are composed of granular modular components (microservices) for high scalability. Hence, it is important to build the applications with a smaller footprint, so that they don't consume too much RAM and CPU. As we move to cloud-native deployments, it's even more important, as we have *pay-per-use* on the cloud. The smaller the footprint, the more we can run with fewer resources on the cloud. This has a direct impact on **Total Cost of Ownership** (**TCO**), one of the key business KPIs.

A smaller footprint also helps us to make changes and deploy them rapidly and continuously. This is very important in the agile world, where the systems are built to embrace change. As businesses change rapidly, applications are also required to embrace changes rapidly to support business decisions. In traditional monolith architectures, even a small change requires an overall build, test, and deployment. In modern architectures, we need flexibility to roll out changes in the functionality in a modular way, without bringing the production systems down.

We have new engineering practices such as A/B testing, where we perform the testing of these functional modules (microservices) in parallel with the older version, to decide whether the new version is good enough to roll out. We perform canary deployments (rolling updates), where the application components are updated, without stopping the production systems. We will cover these architectural requirements in more detail in the *DevOps – continuous integration and delivery* section later in this chapter.

Quicker bootstrap

Scalability is one of the most important requirements. Modern applications are built to scale up and down rapidly based on the load. The load has increased exponentially and modern-day applications are required to handle any load gracefully. With a smaller footprint, it's also expected that these application components (microservices) boot up quickly to start handling the load. As we move toward more serverless architectures, the application components are expected to handle bootup and shutdown on request. This requires a very rapid bootup strategy.

A quicker bootstrap and smaller footprint also pose the challenge of building application components with embeddable VM. The container-based approach requires these application components to be immutable.

Polyglot and interoperability

Polyglot is the reality: each language has its own strengths and will continue to have, so we need to embrace this fact. If you look at the core logic of the interpreter/compiler, they are all the same. They all try to achieve similar levels of optimization and generate the fastest running machine code with the smallest footprint. What we need is an optimum platform that can run these various applications, written in different languages, and also allow interoperability between them.

With these architecture requirement lists in mind, let's now understand how GraalVM works and how it addresses these requirements.

Learning what the GraalVM architecture is

GraalVM provides a Graal JIT compiler, an implementation of JVMCI (which we covered in the previous chapter), which is completely built on Java and uses C2 compiler optimization techniques as the baseline and builds on top of it. Graal JIT is much more sophisticated than a C2 compiler. GraalVM is a drop-in replacement for JDK, which means that all the applications that are currently running on JDK should run on GraalVM without any application code changes.

While GraalVM is built on Java, it not only supports Java, but also enables Polyglot development with JavaScript, Python, R, Ruby, C, and C++. It provides an extensible framework called **Truffle** that allows any language to be built and run on the platform.

GraalVM also provides AOT compilation to build native images with static linking. GraalVM comes with the following list of runtimes, libraries, and tools/utilities (this is for the GraalVM 20.3.0 version. The latest list of components can be found at https://www.graalvm.org/docs/introduction/.)

First, let's have a look at the core components in the following table:

Runtimes	
Java Hotspot VM	Java Hotspot VM, with Interpreter, C1 compiler, and Graal JIT compiler (replacing the C2 compiler).
JavaScript and Node.js Environment	JavaScript and Node.js runtimes that are built on GraalVM. This runtime internally uses Graal JIT to run faster.
LLVM	LLVM runtime to execute programming languages that can be transformed to LLVM bitcode.
Libraries	
GraalVM JIT compiler	GraalVM compiler – a dynamic JIT compiler that improves the efficiency and speed of applications through a unique approach to code analysis and optimization. This topic will be covered in detail in the next chapter.
JavaScript Interpreter	JavaScript Interpreter is an ECMAScript-compliant JavaScript engine that is used to run JavaScript code.
LLVM Bitcode Interpreter	LLVM Bitcode interpreter is an implementation of the lli tool to execute programs directly from LLVM bitcode. We will be covering the LLVM interpreter in more detail later in the chapter in the *Sulong – LLVM* section.
GraalVM Polyglot API	GraalVM Polyglot API – the APIs for combining programming languages in a shared runtime. We will be going through this in detail in *Chapter 6, Truffle*, and *Chapter 7, Graal Polyglot (Java, Node)*.
Utilities/Tools	
JavaScript REPL with the GraalVM JavaScript interpreter	This is the REPL console for JavaScript. REPL is the Read-Evaluate-Print-Loop console, where the programmer can run JavaScript code snippets on an interactive console.
LLVM bitcode interpreter command-line utility	This command-line utility is used to perform bitcode interpretation on native language code
GraalVM Updater	GraalVM Updater (gu) is a command-line utility used to install and manage the additional/optional packages/runtimes/utilities that come with Graal. This is used to install any new releases of interpreters/runtimes.

Next, let's go through the list of additional tools and utilities in the following table:

Native Image	This is a set of tools to help build native images. Graal provides AOT compilation to build native images. We will be covering Graal AOT compilation in more detail and building a native image later in the chapter in the *Substrate VM (Graal AOT and native image)* section.
LLVM toolchain (Sulong)	LLVM (also referred to as Sulong) is a toolchain and APIs for compiling native programs to bitcode. It also allows us to run LLVM programs on GraalVM. We will be going into much more detail on LLVM later in the chapter in the *Sulong – LLVM* section.
Python Interpreter	The Python interpreter is an implementation of the Python 3.8.5 language.
Ruby Interpreter	The Ruby interpreter is an implementation of the Ruby 2.7.3 programming language.
R Interpreter	The R interpreter is a GNU R 3.6.1 implementation of the R programing language.
GraalWasm	GraalWasm is an implementation of the WebAssembly programming language. Wasm is a portable runtime for web applications for both clients and servers. It comes with all major browsers.

Now that we are aware of the components in GraalVM, we will go through the various editions of GraalVM that are available, and the differences between these editions.

Reviewing the GraalVM editions (Community and Enterprise)

GraalVM is available as Community and Enterprise Editions:

- **Community Edition**: GraalVM **Community Edition** (**CE**) is an open source edition built as an OpenJDK distribution. Most of the components of GraalVM are GPL 2, with a classpath exception licensed. For more details on licensing, please refer to `https://github.com/oracle/graal#license`. GraalVM CE is based on OpenJDK 1.8.272 and OpenJDK 11.0.9. GraalVM CE is community supported. It can be deployed in production. However, it does not come with the required support services from Oracle. Oracle also provides a Docker image, which is readily downloadable, for testing and evaluation (refer to `https://www.graalvm.org/docs/getting-started/container-images/` for further details).

- **Enterprise Edition**: GraalVM **Enterprise Edition** (**EE**) is a licensed version under the GraalVM OTN license agreement. This is free for evaluation and building non-production applications. GraalVM EE provides additional performance (~20% faster than CE and dynamic languages such as JavaScript, Python, R, and Ruby are ~2x faster), a smaller footprint (~2x smaller than CE), security (native code memory protection), and scalability for running production enterprise applications. EE comes with additional debugging tools, such as Ideal Graph Visualizer, which helps in not only debugging performance issues, but also in fine-tuning the applications for best performance on GraalVM. GraalVM EE comes with support services. For Oracle cloud customers, GraalVM EE support is available as part of the subscription. GraalVM EE also has a managed mode, which does better heap management, avoiding page faults and crashes. GraalVM EE is available for clients who already have a Java SE subscription. The benefit of EE is that there are patented optimizations that provide additional performance boost that varies depending on the workload, Profile Guided Optimization and a better G1GC garbage collector for an additional performance boost.

For Oracle cloud customers, GraalVM EE support is available as part of the subscription. GraalVM EE also has a managed mode, which does better heap management, avoiding page faults and crashes. GraalVM EE is available for clients who already have a Java SE subscription. GraalVM Enterprise is available as a download and available for free use in Oracle Cloud instance. GraalVM Enterprise is available on Oracle Cloud Ampere 1 compute instances with much lower Cost/Performance than running on an x86 cloud instance. Now that we know the various available editions of GraalVM, and what runtimes, tools, and frameworks come with it, let's get into the details of the GraalVM architecture.

Understanding the GraalVM architecture

In this section, we will look at the various architectural components of GraalVM. We will look at how various runtimes, tools, and frameworks come together to provide the most advanced VM and runtime. The following diagram shows the high-level architecture of GraalVM:

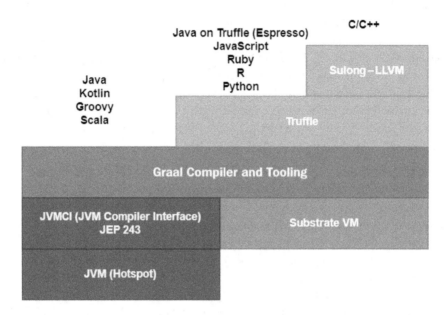

Figure 3.1 – Graal VM architecture

Let's go through each of these components in detail.

JVM (HotSpot)

JVM HotSpot is the regular Java HotSpot VM. The C2 compiler, which is part of the HotSpot VM, is replaced with the Graal JIT compiler implementation. The Graal JIT compiler is an implementation of **Java Virtual Machine Compiler Interface** (**JVMCI**) and plugs into the Java VM. We covered the architecture of JVM HotSpot in the previous chapters. Please refer to them for a more in-depth understanding of how JVM HotSpot works and the various architectural components of JVM.

Java Virtual Machine Compiler Interface (JVMCI)

JVMCI was introduced in Java 9. This allowed compilers to be written as plugins that JVM can call for dynamic compilation. It provides an API and a protocol to build compilers with custom implementations and optimizations.

The word *compiler* in this context means a just-in-time compiler. We went into a lot of detail on JIT compilers in the previous chapters. GraalVM uses JVMCI to get access to the JVM objects, interact with JVM, and install the machine code into the code cache.

Graal JIT implementation comes in two modes:

- `libgraal`: `libgraal` is an AOT compiled binary that is loaded by HotSpot VM as a native binary. This is the default mode and the recommended way to run GraalVM with HotSpot VM. In this mode, `libgraal` uses its own memory space and does not use the HotSpot heap. This mode of Graal JIT has quick bootup and improved performance.

- `jargraal`: In this mode, Graal JIT is loaded like any other Java class, and hence it goes through a warm-up phase and runs with an interpreter until the hot methods are identified and optimized. This mode can be invoked by passing the `--XX:-UseJVMCINativeLibrary` flag from the command line.

In OpenJDK 9+, 10+, and 11+, we use the `-XX:+UnlockExperimentalVMOptions`, `-XX:+UseJVMCICompiler`, and `XX:+EnableJVMCI` flags to run the Graal compiler, instead of the C2 compiler. GraalVM, by default, uses the Graal JIT compiler. It is always advisable to use GraalVM distributions, as these have the latest changes. OpenJDK gets the changes merged at a slower rate.

In the next chapter, we will be going into detail on how Graal JIT is better than the C2 JIT, using a sample code. We will be using the debugging tools and utilities that come with Graal to demonstrate the optimizations that Graal JIT performs at runtime.

Graal compiler and tooling

The Graal compiler is built on JVMCI and provides a better JIT compiler (C2 as we covered in both the previous chapters) implementation, with further optimizations. The Graal compiler also provides an AOT (Graal AOT) compilation option to build native images that can run standalone with embedded VMs.

Graal JIT compiler

We looked at the JVM architecture in *Chapter 1, Evolution of Java Virtual Machine.* For reference, here is the high-level architecture overview of JVM:

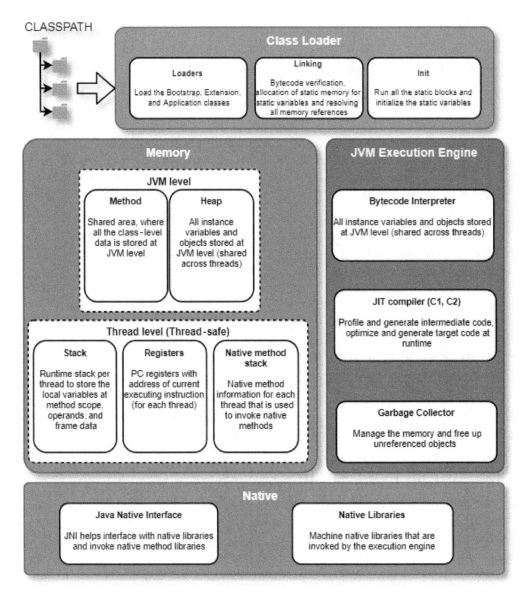

Figure 3.2 – JVM architecture with a C2 compiler

As you can see, the C1 and C2 compilers implement the JIT compilation as part of the JVM execution engine. We went into a lot of detail on how C1 and C2 optimize/deoptimize the code based on the compilation threshold.

GraalVM replaces the JIT compiler in JVM and incorporates further optimization. The following diagram shows the high-level architecture of GraalVM:

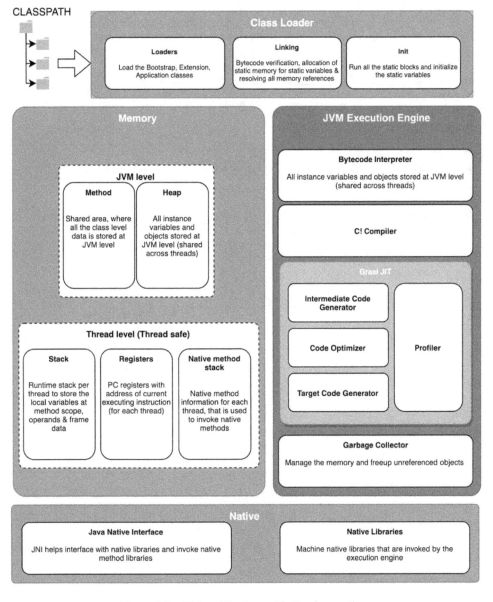

Figure 3.3 – VM architecture with Graal compiler

One of the differences between the JVM JIT compiler and Graal JIT is that Graal JIT is built to optimize the intermediate code representation (**abstract syntax tree (AST)** and using Graal graphs, or Graal intermediate representation). Java represents the code as an AST, an intermediate representation, while compiling.

Any language expressions and instructions can be converted and represented as ASTs; this helps in abstracting the language-specific syntax and semantics from the logic of optimizing the code. This approach makes GraalVM capable of optimizing and running code written in any language, as long as the code can be converted into an AST. We will be doing a deep dive into Graal graphs and ASTs in *Chapter 4, Graal Just-In-Time Compiler*.

The four key components of the Graal JIT compiler are as follows:

- **Profiler**: As the name suggests, it profiles the running code and generates the information that is used by the code optimizer to take decisions or make assumptions regarding optimization.

- **Intermediate Code Generator**: This generates the intermediate code representation, which is the input for the code optimizer.

- **Code Optimizer**: This uses the data that is collected by profiles and optimizes the intermediate code.

- **Target Code Generator**: The optimized code is then converted to the target machine code.

The following diagram shows how Graal JIT works at a very high level:

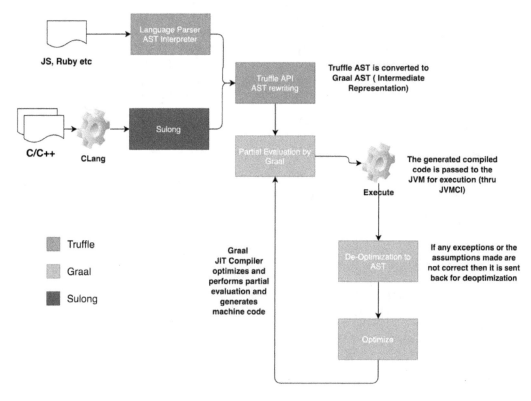

Figure 3.4 – Graal JIT compilation – a high-level flowchart

Let's understand this flowchart better:

- The JVM language (Java, Kotlin, Groovy, and so on) code runs on Graal JIT natively, and Graal JIT optimizes the code.

- The non-JVM languages (such as JavaScript and Ruby) implement a language parser and interpreter using the Truffle API. The language interpreters convert the code into AST representation. Graal runs the JIT compilation on the intermediate representation. This helps in leveraging all the advanced optimization techniques implemented by Graal to non-JVM languages.

- The native LLVM-based languages (such as C/C++, Swift, and Objective C) follow a slightly different route to convert to the intermediate representation. Graal Sulong is used to create the intermediate representation that is used by Graal. We will be talking about Truffle and Sulong later in this chapter.

Graal JIT optimization strategies

Graal JIT optimization strategies are built from the ground up, based on the best practices from the C2 JIT compiler optimization strategies. Graal JIT builds on top of the C2 optimization strategies and provides more advanced optimization strategies. Here are some of the optimization strategies that the Graal JIT compiler performs:

- Partial escape analysis

- Improved inlining (`http://aleksandar-prokopec.com/resources/docs/prio-inliner-final.pdf`)

- Guard optimization (`http://lafo.ssw.uni-linz.ac.at/papers/2013_VMIL_GraalIR.pdf`)

- Chaining lambdas

- Inter-procedural optimization

We will be covering these optimization strategies in detail, with the help of sample code and examples, in the next chapter.

Truffle

The Truffle framework is an open source library for building the interpreters and the tools/utilities (such as integrated development environments, debuggers, and profilers). The Truffle API is used to build language interpreters that can run on GraalVM, leveraging the optimizations provided by GraalVM.

The Graal and Truffle frameworks consist of the following APIs that enable Polyglot:

- **Language Implementation Framework**: This framework is used by the language implementers. It also comes with a reference implementation of a language called **SimpleLanguage**. We will be going through this in detail in *Chapter 9, Graal Polyglot – LLVM, Ruby, and WASM*.

- **Polyglot API**: This set of APIs aids interaction between code written in different languages (guest languages) with Java (the host language). For example, a Java (host) program can embed R (guest) language code to perform some machine learning/AI logic. The Polyglot API provides the framework that will help the language programmers to manage the objects between the guest and the host.

- **Instrumentation**: The Truffle Instrumentation API provides the framework for utilities/tool builders to build integrated development/debugging environments, tools, and utilities. The tools and utilities that are built using the Truffle Instrumentation API can work with any language that is implemented with Truffle. This provides a consistent developer experience across various languages and leverages the sophisticated debugging/diagnostic capabilities of JVM.

Figure 3.5 shows the high-level architecture of how Truffle acts as an intermediate layer between GraalVM and other language interpreters. The individual language interpreters are implemented using the Truffle API. Truffle also provides an interoperability API, for calling methods and passing data between methods across various language implementations:

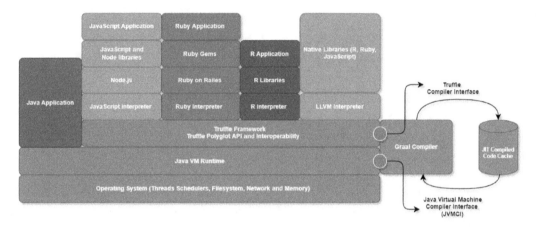

Figure 3.5 – Truffle architecture

As represented in the previous diagram, Java applications run directly on the GraalVM, with Graal Compiler replacing the C2 JIT compiler. Other language programs run on top of the Truffle Language Implementation framework. The respective language interpreters use the Truffle to implement the interpreters. Truffle combines the code along with the interpreter to produce the machine code, using partial evaluation.

AST is the intermediate representation. AST provides the optimum way to represent the syntactic structure of the language, where typically, the parent node is the operator, and the children node represents the operands or operators (based on cardinality). The following diagram shows a rough representation of AST:

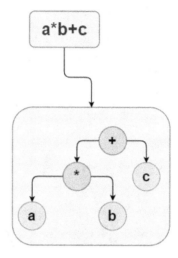

Figure 3.6 – AST for simple expression

In this diagram, **a**, **b**, and **c** can be any variables (for loosely typed languages). The interpreter starts assuming "generics" based on the profiling of various executions. It then starts assuming the specifics and will then optimize the code using partial evaluation.

Truffle (language interpreters written with Truffle) runs as an interpreter and Graal JIT kicks in to start identifying optimizations in the code.

The optimizations are based on speculation, and eventually, if the speculation is proven to be incorrect at runtime, the JIT will re-optimize and recompile (as shown in the previous diagram). Re-optimization and recompiling is an expensive task.

Partial evaluation creates an intermediate representation of the language, from the code and the data, and as it learns, and identifies new data types, it deoptimizes to the AST interpreter, applies optimizations, and does the node rewriting and recompiles. After a certain point, it will have the optimum representation. The following diagram explains how Truffle and Graal optimize intermediate representation:

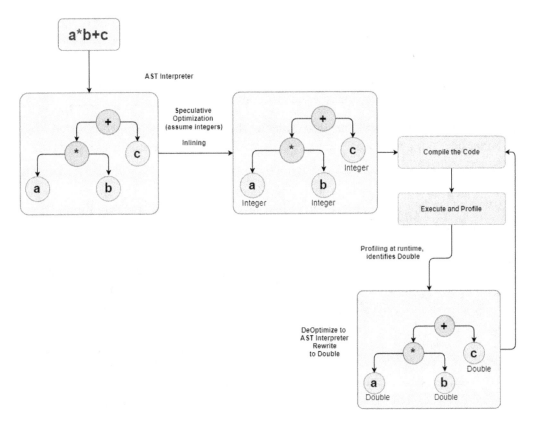

Figure 3.7 – AST optimization by Graal JIT

Let's understand this diagram better:

- The expression is reduced to an AST. In the AST nodes, the leaf nodes are operands. In this example, we have taken a very simple expression to understand how partial evaluation works. In a language such as JavaScript, which is not a strongly typed language, a, b, and c can be any data type (sometimes referred to as generics). Evaluating a generic in an expression is a costly operation.

- Based on the profiling, Graal JIT speculates and assumes a specific data type (in this example, as an integer), optimizes the code to evaluate the expression for integers, and compiles the code.

- In this example, it is using an inlining optimization strategy. The Graal JIT compiler has various other optimization strategies that are applied based on the use case.

- When, during runtime, the compiler identifies a control flow where one of the operands is not really an integer, it deoptimizes and rewrites the AST with the new data type and optimizes the code.

- Following a few iterations of running this optimization/deoptimization, the compiler will eventually generate the most optimum code.

The key difference here is that Graal is working on the AST and generating optimized code, and it does not matter what language is used to write the source code as long as the code is represented as AST.

The following diagram shows a high-level flow of how different languages run on GraalVM, with Truffle acting as an intermediate layer, to execute any programming language code on GraalVM:

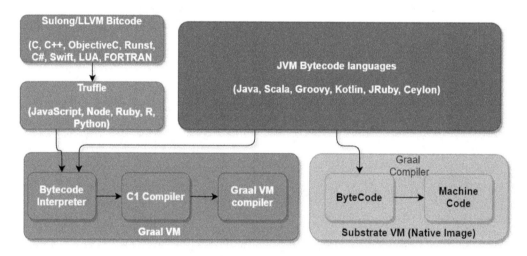

Figure 3.8 – Truffle and Graal compilation flowchart

This diagram illustrates a simpler representation of how Truffle acts as a layer between non-JVM languages and GraalVM. The code can also be built directly as a native image with Substrate VM.

Truffle API is used along with a custom annotations processor to generate interpreter source code, which is then compiled. Java code does not need the intermediate representation. It can be compiled directly to run on GraalVM. We will discuss Truffle interpreters and how to write a custom interpreter in *Chapter 9, Graal Polyglot – LLVM, Ruby, and WASM*. We will cover the Truffle Polyglot API in *Chapter 6, Truffle for Multi-language (Polyglot) support*.

Truffle also provides a framework called the **Truffle Instrument API** for building tools. Instruments provide fine-grained VM-level runtime events, which can be used to build profiling, tracing analyzing, and debugging tools. The best part is that the language interpreters built with Truffle can use the ecosystem of Truffle instruments (for example, VisualVM, Chrome Debugger, and GraalVM Visual Studio Code Extension).

Truffle provides the **Polyglot Interoperability Protocol**. This protocol defines the message that each language needs to implement and supports the passing of data between the Polyglot applications.

Sulong – LLVM

LLVM is an open source project that is a collection of modular, reusable compilers and toolchains. There are a lot of language (C, C++, Fortran, Rust, Swift, and so on) compilers that are built on LLVM, where LLVM provides the intermediate representation (also known as LLVM-IR).

The Sulong pipeline looks different from what we looked at in other language compilers that are running on Truffle. The following diagram shows how C/C++ code gets compiled:

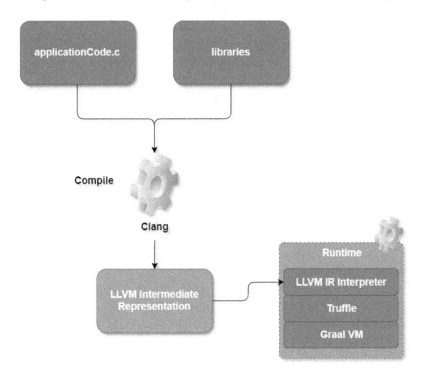

Figure 3.9 – LLVM compilation flowchart

This diagram shows how an application code written in C is compiled and run on GraalVM. The application code in native languages such as C/C++ is compiled in Clang, into an intermediate representation. This LLVM intermediate representation runs on the LLVM intermediate representation interpreter, which is built on the Truffle API. Graal JIT will further optimize the code at runtime.

SubstrateVM (Graal AOT and native image)

Applications on Graal can be deployed on GraalVM or SubstrateVM. SubstrateVM is embeddable VM code that gets packaged during AOT compilation in native images.

Graal AOT compilation is a very powerful way to create native binaries for a particular targeted OS/architecture. For cloud-native workloads and serverless, this is a very powerful option for achieving a smaller footprint, faster startups, and, more importantly, embeddable runtimes (providing immutability).

Rapid componentized modular deployment (containers) also poses management and versioning challenges. This is typically called **Configuration Drift**, which is one of the major issues that we face in managing a large number of containers in high-availability environments. Typically, container infrastructure is built by a team and, over time, it is managed by different teams. There are always situations where we are forced to change the configuration of the VMs/containers/OS in a particular environment that we may never trace. This causes a gap between production and the DR/HA environment.

Immutable infrastructure (images) helps us do better version control of the infrastructure. It also gives us more confidence in testing, as the underlying infrastructure on which our application containers are running is immutable, and we are certain about the test results. To build immutable components, we require an embeddable VM (with a small footprint). SubstrateVM provides that embeddable VM.

In AOT compilation, the code is compiled directly to the machine code and executed. There is no runtime profiling or optimization/deoptimization. The Graal AOT compiler (also referred to as the "native image" compiler) performs static analysis and static initializations on the code and produces a VM-embedded executable code. The optimization performed by AOT is based on the reachability of the code. The following diagram shows the compilation process:

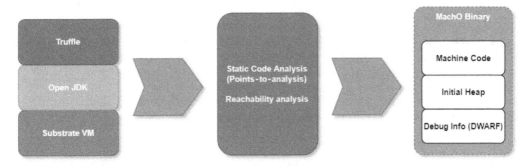

Figure 3.10 – Graal AOT compilation

This diagram shows how the Graal AOT compiles native images and embeds SubstrateVM as part of the native image. One of the disadvantages of AOT compilation is that the VM cannot optimize the code based on runtime profiling, as in JIT. To address this issue, we use a profile guided optimization strategy to capture the runtime metrics of the application and use that profiled data to optimize the native image by recompiling.

Profile Guided Optimization (PGO)

GraalVM uses **Profile Guided Optimization** (**PGO**) to optimize native images based on the runtime profiling data. This is one of the features that is available in Enterprise Edition only. The following diagram shows how a PGO pipeline works:

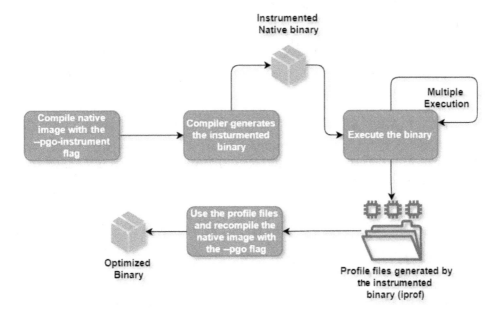

Figure 3.11 – Graal AOT compilation with PGO

Let's understand this workflow better:

- When the code is compiled with `native-image`, we use the `--pgo-instrumented` flag. This will tell the compiler to inject instrumentation code into the compiled code.

- When we start running this instrumented native image, the profiler starts collecting the runtime data and then starts creating the profile files (`.ipof`).

- Once we have run the native image with various workloads (all possible workloads – to capture as much instrumentation as possible), we can then recompile the native image with the `--pgo` flag (`native-image --pgo=profile.iprof`), providing the profile files as input. The Graal native image compiler creates the optimum native image.

We will be building a native image with profile-guided optimization in the next chapter with the help of real examples and also understand how memory management works in native images.

GraalVM is a great runtime for the modern microservices architecture. In the next section, we will go through the various features of GraalVM that help to build a microservices application.

An overview of GraalVM microservices architecture

GraalVM is ideal for microservices architecture. One of the most important requirements for certain microservices architecture is a smaller footprint and faster startup. GraalVM is an ideal runtime for running Polyglot workloads in the cloud. Cloud-native frameworks are already available on the market that can build applications to run optimally on GraalVM, such as Quarkus, Micronut, Helidon, and Spring. These frameworks are found to perform almost 50x times faster when running as native images. We will go into detail about how GraalVM is the right runtime and platform for microservices in *Chapter 10, Microservices Architecture with GraalVM*.

Understanding how GraalVM addresses various non-functional aspects

In this section, we will go through the typical non-functional requirements of a microservices cloud-native architecture, and how GraalVM addresses these requirements.

Performance and scalability

Performance and scalability are among the more important non-functional requirements of a microservices cloud-native architecture. The microservices are automatically scaled out and down by orchestrators such as Kubernetes. This requires the microservices to be built on a runtime that starts up quickly and runs fast, consuming minimal cloud resources. GraalVM AOT compilation helps to build native images that perform on a par with native languages such as C/C++.

To understand how AOT compiled code (native image) is faster than JIT compiled code, let's look at the steps that JIT and AOT follow at runtime:

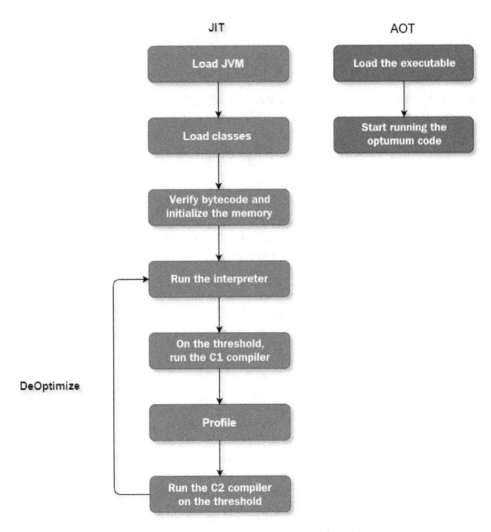

Figure 3.12 – Graal JIT versus the AOT flowchart

This diagram shows the high-level steps for JIT and AOT. JIT optimizes the code over a period of time by profiling the code at runtime. There are performance overheads, as there is additional profiling, optimizing, and deoptimizing that is done by the JVM at runtime.

It is observed, based on the Apache Bench benchmark, that while GraalVM JIT throughput and performance is lower than AOT at the beginning, as the number of requests increases, Graal JIT optimizes and performs better than Graal AOT after around 14,000 requests per second.

It is also observed that Graal AOT performs 50 times faster than Graal JIT and has a 5x smaller memory footprint than Graal JIT.

Graal AOT with PGO throughput is consistent and sometimes better than Graal JIT. However, for long-running tasks, Graal JIT might have better throughput. So, for the best throughput and consistent performance, Graal AOT with PGO is the best.

Please refer to the benchmark study published at `https://www.infoq.com/presentations/graalvm-performance/` and `https://www.graalvm.org/why-graalvm/`.

There are further benchmark studies that are published with academic collaborators at `https://renaissance.dev`.

Here's what we can conclude:

- GraalVM Native Image (AOT) is best for faster startups and applications that require a smaller footprint, such as serverless applications and container microservices.

- GraalVM JIT is best for peak throughputs. Throughput is a very important aspect for long-running processes, where scalability is critical. This could be high-volume web application servers such as e-commerce servers and stock market applications.

- A combination of garbage collection configuration and JIT will help in achieving reduced latency. Latency is very important as regards the responsiveness of applications. When we are running high throughput, it's possible that on occasion, garbage collection slows down the response.

There is not a hard and fast rule for using it. It depends on various combinations that we need to decide between JIT and AOT, and various other configurations that are possible. We will explore various compiler and native image configurations in the next chapter.

Security

GraalVM security is built on JVM security, which is based on the sandbox model. Let's have a very quick review of how the sandbox model works:

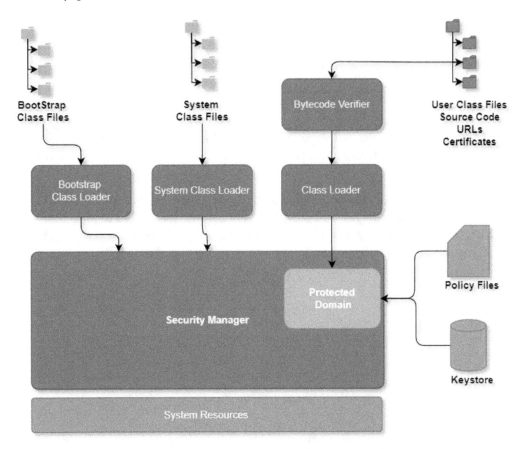

Figure 3.13 – JVM security model

In Java 2 security architecture, all the class files are verified by bytecode verifier (please refer to both the previous chapters for more details on class loaders). The bytecode verifier checks whether the class files are valid and look for any overflows, underflows, data type conversions, method calls, references to classes, and so on.

Once the bytecode is verified, the dependent classes are loaded by the class loader. Please refer to *Chapter 1, Evolution of Java Virtual Machine*, to understand how the class loader subsystem works. Class loaders work with Security Manager and access control to enforce security rules that are defined in the policy files. Java code that is downloaded over the network is checked for a signature (represented as `java.security.CodeSource`, including the public key).

Security Manager (`java.lang.SecurityManager`) is the most important component for handling authorizations. Security Manager has various checks in place to ensure that the authorization is done. The access controller (`java.security.AccessContoller`) class is another critical class that helps control access to system resources.

Keystore is a password-protected store that holds all the private keys and certificates. Each entry in the store can also be password-protected.

Java Security is extendable, with custom security implementations called **Security Providers**.

GraalVM builds on the Java security model and abstracts it to address enforcing security at intermediate representation level. GraalVM does not recommend running untrusted code on Security Manager.

The GraalVM security model uses the Truffle language implementation framework API for JVM host applications to create an execution context, which is passed to the guest application (application code written in different languages). The following diagram shows the high-level architecture of how GraalVM allows the guest and host applications to interoperate and determine how access is controlled:

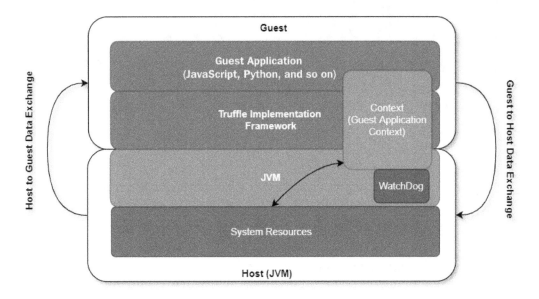

Figure 3.14 – Graal security model

The execution context (org.graalvm.polyglot.Context) defines the access control for the guest applications. Based on the access control that is defined in the execution context, the guest applications get access to the system's resources. GraalVM provides a Polyglot API to create these access controls, with an execution context, to set the access privileges to access various functions, such as File I/O, Threading, and Native Access. Based on what privileges are set by the host, the guest will have that access. A watchdog thread is used to timebound the context. The watchdog will close the context, in the given time, to free up the resources, and restrict access based on time.

The following code demonstrates how the execution context can be set:

```
Context context = Context.newBuilder().allowIO(true).build();
Context context = Context.newBuilder()
                .fileSystem(FileSystem fs).build();
Context context = Context.newBuilder()
                .allowCreateThread(true).build()
Context context = Context.newBuilder()
                .allowNativeAccess(true).build()
```

GraalVM also offers an API to exchange objects between host and guest applications:

- **Guest to Host Data Exchange**: The guest application can call the host methods and may pass the data. However, this is controlled based on method access modifiers and the host access policy (ALL, NONE or EXPLICIT – @HostAccess.Export Annotation, for example).

- **Host to Guest Data Exchange**: The objects passed from the host to the guest need to be handled by guest languages. The data is passed through the context, for example:

```
Value a = Context.create().eval("js", "21 + 21");
```

Value a can be returned by the host application to the JavaScript guest application with a value of 42 (after evaluation).

We will be covering this in detail in *Chapter 6, Truffle for Multi-language (Polyglot) support*, with the help of a real example.

GraalVM EE also provides a managed execution mode for LLVM intermediate representation code to handle any memory violations and faults. Please refer to https://docs.oracle.com/en/graalvm/enterprise/19/guide/security/security-guide.html for more details.

DevOps – continuous integration and delivery

DevOps automation is one of the core requirements of any modern, cloud-native architecture. GraalVM integrates very well into the DevOps pipeline. The following diagram illustrates a typical GitOps pipeline with representative software (GraalVM integrates into any stack of the DevOps software stack):

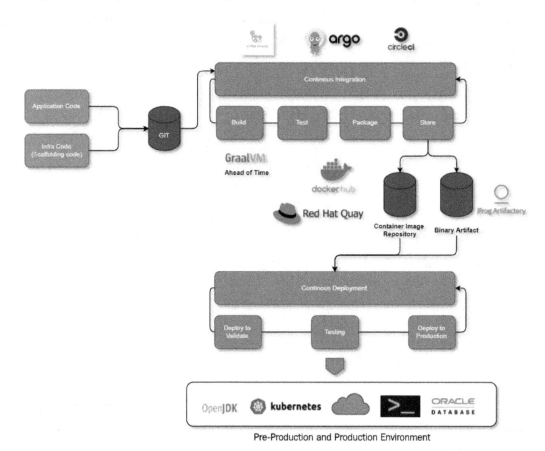

Figure 3.15 – GitOps with GraalVM

Let's understand this diagram better.

The **Continuous Integration (CI)** pipeline gets triggered by a typical pull request from the Git repository with changes in the application code and infrastructure code. CI tools such as GitHub actions, Argo CD, or CicleCI can be used to orchestrate a CI pipeline. A typical CI pipeline consists of the following:

- **Build**: In the build phase, the release tagged code is pulled from the appropriate branch from the Git repository. The code is verified (any static code analysis) and built. For cloud-native, typically, the code is built as native images, using the Graal AOT compiler.

- **Test**: The code is tested with unit testing scripts and further verified for any security vulnerabilities.

- **Package**: Once the code passes all the tests, the code is typically packaged into the cloud-native target runtime (using a Docker image or VM or any other binary format). The target could be a serverless container or Docker container or a VM.

- **Store**: The final binaries are stored on binary stores or repositories, such as Docker Hub or Red Hat Quay (if it's a Docker image).

The Continuous Deployment pipeline can either be triggered based on a release plan or can be manually triggered (depending on the release plan and strategy). Continuous Deployment typically has the following phases:

- **Deployment for Validation**: The final binary ID deployed to an environment where the binary can now be tested end to end. Various strategies can be followed:

    a. **Traditionally**: We have an Integration Test Environment and a User Acceptance Test Environment (or Pre-Production Environment) for various levels of validation and testing.

    b. **Blue/Green Deployment**: There are two parallel environments (called Blue and Green). One of them will be in production, let's assume Blue. The Green environment can be used to test and validate our code. Once we are sure that the new release is working fine, we use the router to switch to the Green environment and use the Blue environment for testing future releases. This provides a high availability way to deploy applications.

c. **Canary Deployments and Rolling Updates**: Canary deployment is more a recent approach of using the same environment for both production and validation. This is a great feature to test our code and compare the new release with the current release (A/B testing). Canary deployments provide an API management layer, which can be used to redirect the traffic to specific endpoints, based on various parameters (such as testing users or a user from a specific department can access the new version, while end users are still using the old version). The application can be deployed on a specific number of servers/nodes (by % or number). As we get more confident with the new release, we can perform rolling updates by increasing the number of nodes, where the new release should run, and open up to a wider circle of users. This also gives the flexibility to perform a phased rollout of new releases (by region or user demography or any parameter).

- **Testing**: There are various levels of testing that are performed, both functional and non-functional. Most of this is performed with automation, and that is also choreographed by the Continuous Delivery pipeline.

- **Production Deployment**: Once it's all tested, the final application is deployed to the production environment. Once again, this deployment may use one of the Traditional or Blue/Green or Canary strategies.

GraalVM provides a very flexible way to deploy the application as a standalone, container, cloud, VM, and Oracle database. There are very sophisticated microservices frameworks, such as Quarkus, Micronaut, and Fn project, that provide native support for GraalVM and integrate very well with modern GitOps tools.

Summary

In this chapter, we explored the GraalVM architecture. Graal JIT is the new implementation of the JIT compiler, which replaces the C2 compiler, and brings in a lot more optimizations. Graal JIT is implemented completely in Java. Truffle provides the interpreter implementation framework and Polyglot framework to get other non-JVM languages into GraalVM.

This chapter provided a good understanding of the various runtimes, frameworks, tools, Graal updater, and utilities that are shipped with GraalVM. We also looked at the two available editions of GraalVM and what the key differences are between these two editions. We went through all the various components of the GraalVM architecture. We also explored some of the non-functional aspects of the architecture, including security model, performance, and DevOps. This is very important if you want to understand how GraalVM can be used to build cloud-native microservices and high-performing applications across various languages.

In the next chapter, we will dig deeper into how Graal JIT works, how we can use the various tools that come with Graal to understand the internal workings of Graal JIT, and how we can use these tools to debug and fine-tune our code.

Questions

1. What are the various editions of GraalVM?

2. What is JVMCI?

3. What is Graal JIT?

4. What is Graal AOT? How does PGO help AOT compilation?

5. What is Truffle? How does it help to run multiple language codes on GraalVM?

6. What is SubstrateVM?

7. What is Guest Access Context?

8. Why is GraalVM the ideal runtime for cloud-native microservices?

Further reading

- Lightweight Cloud-Native Java Applications (https://medium.com/graalvm/lightweight-cloud-native-java-applications-35d56bc45673)

- Java on Truffle — Going Fully Metacircular (https://medium.com/graalvm/java-on-truffle-going-fully-metacircular-215531e3f840)

- GraalVM (https://www.graalvm.org/)

- GraalVM Enterprise Edition (https://docs.oracle.com/en/graalvm/enterprise/20/index.html)

- GraalVM Git (https://github.com/oracle/graal)

4
Graal Just-In-Time Compiler

In *Chapter 3, GraalVM Architecture*, we went through the GraalVM architecture and the various components that constitute it. We went through some details of the GraalVM Polyglot architecture with Truffle and touched upon the Graal's **just-in-time** (JIT) compiler. We looked at how Graal JIT plugs into the Java Virtual Machine by implementing the Java Virtual Machine Compiler Interface. In this chapter, we will explore how the Graal JIT compiler works by running sample code and visualizing the Graal graphs and optimizations that Graal JIT performs, using the Ideal Graph Visualizer tool.

In this chapter, we will cover the following topics:

- Setting up the environment
- Understanding the Graal JIT compiler
- Understanding Graal compiler optimizations
- Debugging and monitoring applications

By the end of the chapter, you will have a very clear understanding of how Graal JIT compilation works, understand the various optimization techniques, know how to diagnose and debug performance problems using the Ideal Graph Visualizer, and be able to fine-tune Graal JIT compiler configurations for optimum performance.

Technical requirements

In this chapter, we will take some sample code and use tools to analyze it. The following are some of the tools/runtimes that are required to follow this chapter:

- OpenJDK (`https://openjdk.java.net/`)
- GraalVM (`https://www.graalvm.org/`)
- VisualVM (`https://visualvm.github.io/index.html`)
- The Ideal Graph Visualizer
- There are some sample code snippets, which are available in our Git repository. The code can be downloaded from `https://github.com/PacktPublishing/Supercharge-Your-Applications-with-GraalVM/tree/main/Chapter04`.
- The Code in Action video for this chapter can be found at `https://bit.ly/3fmPsaP`.
- Seafoam (an Open Source project, which provides enhanced functionality and a good alternate tool for IGV)

Setting up the environment

In this chapter, we will be using VisualVM and the Ideal Graph Visualizer to understand how Graal JIT works. This understanding will help us, in the subsequent chapters, to build optimum code with Graal.

Setting up Graal

In *Chapter 3*, *GraalVM Architecture*, we discussed the two editions of Graal – Community Edition and **Enterprise Edition** (**EE**). Graal Community Edition can be downloaded from the Git repository mentioned in the *Technical requirements* section, while EE requires you to register with Oracle to download it. EE is available for free for evaluation and non-production applications.

Installing the Community Edition

To install GraalVM Community Edition, go to `https://github.com/graalvm/graalvm-ce-builds/releases` and download the latest release for the target operating system (macOS, Linux, and Windows). At the time of writing this book, the latest version is 21.0.0.2, with base Java 8 or Java 11 versions. The Community Edition is built on OpenJDK.

Please follow the instructions provided next for your target operating system. The latest instructions can be found at `https://www.graalvm.org/docs/getting-started/#install-graalvm`.

Installing GraalVM on macOS

For macOS, after downloading the GraalVM archive file, unzip the archive and copy the contents of the unzipped folder to `/Library/Java/JavaVirtualMachines/<graalvm>/Contents/Home`.

Once we have copied the files, we have to export the paths to access the GraalVM binaries. Let's run the following `export` commands on the terminal:

```
export PATH=/Library/Java/JavaVirtualMachines/<graalvm>/
Contents/Home/bin:$PATH
export JAVA_HOME=/Library/Java/JavaVirtualMachines/<graalvm>/
Contents/Home
```

For macOS Catalina and later, the `quarantine` attribute needs to be removed. It can be done with the following command:

```
sudo xattr -r -d com.apple.quarantine <graalvm-path>
```

If this is not done, you will see the following error message:

Figure 4.1 – Error message while running Graal on MacOS

SDKMAN provides an automated way of installing GraalVM. Please refer to `https://sdkman.io/` for more details.

Installing GraalVM on Linux

To install GraalVM on Linux, extract the downloaded zip file, copy it to any target folder, and set the PATH and JAVA_HOME paths to point to the folder where the extracted files are. To do this, execute the following commands on the command line:

```
export PATH=<graalvm>/bin:$PATH
export JAVA_HOME=<graalvm>
```

Installing GraalVM on Windows

To install GraalVM on Windows, extract the .zip file, copy it to any target folder, set the PATH and JAVA_HOME paths to point to the folder where the extracted files are. To set the PATH environment variables, execute the following commands on the terminal:

```
setx /M PATH "C:\Progra~1\Java\<graalvm>\bin;%PATH%"
setx /M JAVA_HOME "C:\Progra~1\Java\<graalvm>"
```

To check that the installation and setup are complete, run the `java -version` command on the terminal.

After executing the command, you should see something like the following output (I am using GraalVM EE 21.0.0 on Java 11. You should see the version that you installed):

```
java version "11.0.10" 2021-01-19 LTS
Java(TM) SE Runtime Environment GraalVM EE 21.0.0 (build
11.0.10+8-LTS-jvmci-21.0-b06)
Java HotSpot(TM) 64-Bit Server VM GraalVM EE 21.0.0 (build
11.0.10+8-LTS-jvmci-21.0-b06, mixed mode, sharing)
```

Let's now explore the folder structure of GraalVM installation. In the GraalVM installation folder, you will find the folder structure explained in the following table:

Folder Path	Description
./bin	This folder has all the binaries that come with GraalVM. This includes the Graal versions of JVM binaries, such as javac, java, jar, and appletviewer, and various Graal implementations of language and some tools. Let's quickly go through some of the important files in this folder: • native-image: Ahead of Time compiler • gu: GraalVM updater, used to install and manage optional GaalVM language runtimes and utilities. • node: Node.js runtime implemented on Graal with Truffle • npm: Node Package Manager implementation • graalpython: Python runtime built on Graal • gem: gem package manager for Ruby • truffleruby: Ruby implementation with Truffle on Graal • wasm: WebAssembly implementation
./include	Contains the files that are needed to integrate with Java from other languages
./jre	Java runtime with Graal JIT
./lib	Library files and Java API class libraries
./man	Markdowns of legal notices
./tools	Tools such as Chrome Inspector, Insight, Ideal Graph Visualizer, and Profiler are installed here. We will be covering these tools later in the chapter.

In the previous chapter, we covered in detail the various runtimes, tools, and utilities that come with Graal. Graal Updater is one of the very important tools that is used to install optional runtimes. To check the runtimes that are available, execute gu list. The following screenshot shows the typical output:

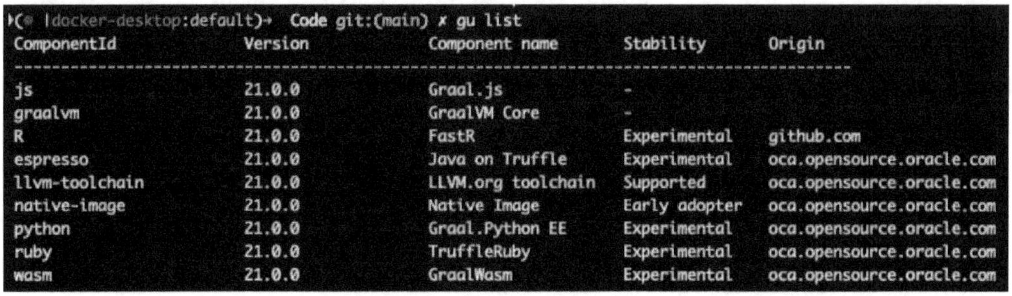

```
▶(● |docker-desktop:default)→  Code git:(main) ✗ gu list
ComponentId          Version         Component name      Stability       Origin
------------------------------------------------------------------------------------------
js                   21.0.0          Graal.js            -
graalvm              21.0.0          GraalVM Core        -
R                    21.0.0          FastR               Experimental    github.com
espresso             21.0.0          Java on Truffle     Experimental    oca.opensource.oracle.com
llvm-toolchain       21.0.0          LLVM.org toolchain  Supported       oca.opensource.oracle.com
native-image         21.0.0          Native Image        Early adopter   oca.opensource.oracle.com
python               21.0.0          Graal.Python EE     Experimental    oca.opensource.oracle.com
ruby                 21.0.0          TruffleRuby         Experimental    oca.opensource.oracle.com
wasm                 21.0.0          GraalWasm           Experimental    oca.opensource.oracle.com
```

Figure 4.2 – Graal Updater listing

We can run gu install <runtime> to install other runtimes.

Installing EE

GraalVM EE is available for free for trial and non-production use. It can be downloaded from https://www.graalvm.org/downloads/.

Select the required GraalVM Enterprise version. The website will redirect you to Oracle's registration page. If you are already registered, you should be able to log in, and you will be redirected to a page from where you can download GraalVM and the supporting tools. At the time of writing this book, the screen looks something like the following screenshot:

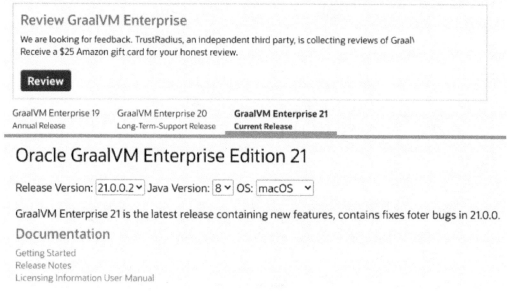

Figure 4.3 – GraalVM EE download page

You can select the right version of EE that you want to download along with the base JDK version. At the time of writing the book, Java 8 and Java 11 are two viable versions. When you scroll down this page, you will find download links for the following:

- **Oracle GraalVM Enterprise Edition Core**: This the code for GraalVM.

- **Oracle GraalVM Enterprise Edition Native Image**: This is the native image tool. It can also be downloaded using Graal Updater later.

- **Ideal Graph Visualizer**: This is a very powerful Graal graph analyzer tool. It needs to be downloaded for this chapter. See the instructions in the *Installing the Ideal Graph Visualizer* section.

- **GraalVM LLVM Toolchain**: This is the LLVM toolchain, which is required if you want to compile and run C/C++ applications on GraalVM.

- **Oracle GraalVM Enterprise Edition Ruby Language Plugin**: This is the Ruby language compiler and runtime. It can also be downloaded using Graal Updater later.

- **Oracle GraalVM Enterprise Edition Python Language Plugin**: This is the Python language compiler and runtime. It can also be downloaded using Graal Updater later.

- **Oracle GraalVM Enterprise Edition WebAssembly Language Plugin**: This is the WebAssembly language compiler and runtime. It can also be downloaded using Graal Updater later.

- **Oracle GraalVM Enterprise Edition Java on Truffle**: This is the JVM implementation on the Truffle interpreter.

Switching between editions

We can have multiple versions/distributions of GraalVM installed on the same machine, and we can switch between these various distributions. In this chapter, we will be switching between the distributions to compare their performance. The best way to switch between distributions is by using Visual Studio Code. Visual Studio Code provides a GraalVM plugin that helps us to add the various distributions, and with single click of a button, allows us to switch between the various distributions. Please refer to `https://www.graalvm.org/tools/vscode/` and `https://marketplace.visualstudio.com/items?itemName=oracle-labs-graalvm.graalvm` for more details. Refer to the *Debugging and monitoring applications* section later in this chapter for more details on how to install Visual Studio Code and use it for debugging applications.

We can also create shell scripts to switch between the various distributions by setting the PATH and JAVA_HOME environment variables to point to appropriate distributions.

Installing Graal VisualVM

Java VisualVM is one of the most powerful tools for analyzing an application's heap, thread, and CPU utilization. VisualVM is widely used to analyze core dumps, heap dumps, and applications that are offline. It is a very sophisticated tool that can identify bottlenecks and optimize Java code.

Since JDK 9, VisualVM has been moved and upgraded to Graal VisualVM. Graal VisualVM extends the functionality to include the analysis of Graal processes, and currently supports JavaScript, Python, Ruby, and R. Graal VisualVM also supports some limited monitoring and analysis functionality for native image processes. Graal VisualVM comes bundled with both Graal Community Edition and EE. Graal VisualVM can be found at .bin/jvisualvm(.exe for windows).

Let's quickly go through the key features of Graal VisualVM. Graal VisualVM has a very intuitive interface. The left panel of the main window (see *Figure 4.3*) shows all the **Local** and **Remote** processes. Using this, we can easily connect to those processes to start our analysis:

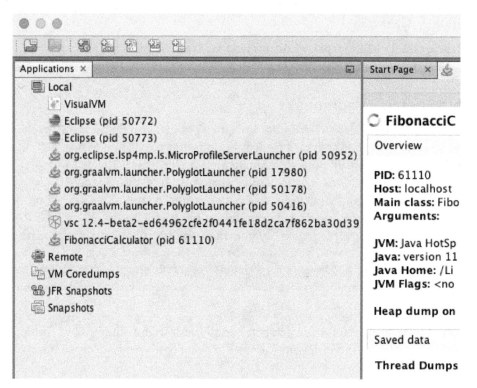

Figure 4.4 – VisualVM, left pane

Once we connect to the process, in the right panel, we will see the following five tabs:

- **Overview**: On this tab, we can see the process configuration, JVM arguments, and system properties. The following screenshot shows the typical screen for the **FibonacciCalculator** process that we are running:

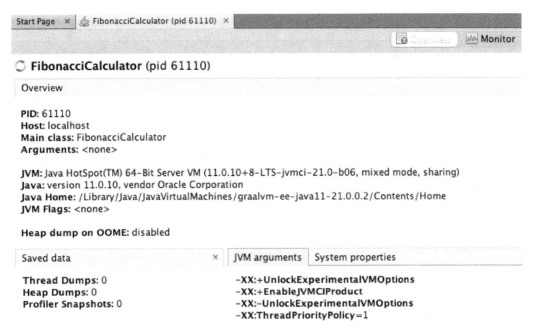

Figure 4.5 – VisualVM – application overview

- **Monitor**: On this tab, we can see CPU usage, heap allocation, the number of classes that are loaded, the number of threads that are running, and so on. We can also force a garbage collection to see how the process behaves. We can perform a heap dump to do a deeper analysis of the heap allocations. Here is a screenshot of the window:

Figure 4.6 – VisualVM – application monitoring

- **Threads**: This tab provides detailed information about the various threads that are running the processes. We can also capture a thread dump to perform further analysis. This tab not only shows the live threads, but we can also analyze the threads that have finished execution. The following screenshot shows the typical **Threads** tab:

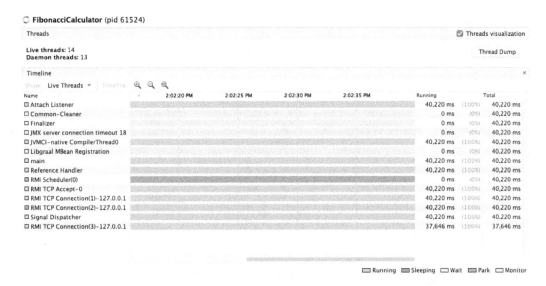

Figure 4.7 – VisualVM – application threads

Here is a typical screenshot of the thread dump that can be used to identify if there are any deadlocks or thread waits:

◯ **FibonacciCalculator** (pid 61524)

Thread Dump

```
2021-05-04 14:02:50
Full thread dump Java HotSpot(TM) 64-Bit Server VM (11.0.10+8-LTS-jvmci-21.0-b06 mixed mode, sharing):

Threads class SMR info:
_java_thread_list=0x00007fc3bba5ef60, length=17, elements={
0x00007fc3dd009000, 0x00007fc3db05f800, 0x00007fc3db93d000, 0x00007fc3db93f800,
0x00007fc3da80b800, 0x00007fc3da88c800, 0x00007fc3db019000, 0x00007fc3db05c800,
0x00007fc3da840000, 0x00007fc3bb00b000, 0x00007fc3dd017000, 0x00007fc3dba5a000,
0x00007fc3dba5b000, 0x00007fc3ba83d000, 0x00007fc3db143800, 0x00007fc3dc84b000,
0x00007fc3db971000
}

"main" #1 prio=5 os_prio=35 cpu=70305.28ms elapsed=71.34s tid=0x00007fc3dd009000 nid=0x1803 runnable  [0x000070000637d000]
   java.lang.Thread.State: RUNNABLE
        at java.util.Formatter.parse(java.base@11.0.10/Formatter.java:2699)
        at java.util.Formatter.format(java.base@11.0.10/Formatter.java:2655)
        at java.io.PrintStream.format(java.base@11.0.10/PrintStream.java:1053)
        - locked <0x00000006ee86cdb8> (a java.io.PrintStream)
        at java.io.PrintStream.printf(java.base@11.0.10/PrintStream.java:949)
        at FibonacciCalculator.main(FibonacciCalculator.java:33)

   Locked ownable synchronizers:
        - None

"Reference Handler" #2 daemon prio=10 os_prio=35 cpu=1.30ms elapsed=71.33s tid=0x00007fc3db05f800 nid=0x3503 waiting on condition  [0x0000700006a92
   java.lang.Thread.State: RUNNABLE
        at java.lang.ref.Reference.waitForReferencePendingList(java.base@11.0.10/Native Method)
        at java.lang.ref.Reference.processPendingReferences(java.base@11.0.10/Reference.java:241)
        at java.lang.ref.Reference$ReferenceHandler.run(java.base@11.0.10/Reference.java:213)

   Locked ownable synchronizers:
        - None

"Finalizer" #3 daemon prio=8 os_prio=35 cpu=0.21ms elapsed=71.33s tid=0x00007fc3db93d000 nid=0x3703 in Object.wait()  [0x0000700006b95000]
   java.lang.Thread.State: WAITING (on object monitor)
        at java.lang.Object.wait(java.base@11.0.10/Native Method)
        - waiting on <0x00000006ee800b08> (a java.lang.ref.ReferenceQueue$Lock)
        at java.lang.ref.ReferenceQueue.remove(java.base@11.0.10/ReferenceQueue.java:155)
        - waiting to re-lock in wait() <0x00000006ee800b08> (a java.lang.ref.ReferenceQueue$Lock)
        at java.lang.ref.ReferenceQueue.remove(java.base@11.0.10/ReferenceQueue.java:176)
        at java.lang.ref.Finalizer$FinalizerThread.run(java.base@11.0.10/Finalizer.java:170)

   Locked ownable synchronizers:
        - None
```

Figure 4.8 – VisualVM – thread dump

- **Sampler**: This tab can be used to take a snapshot of the running process and carry out analysis on CPU, memory, and so on. Here is a screenshot that shows the memory usage for the snapshot we take by clicking the **Snapshots** button:

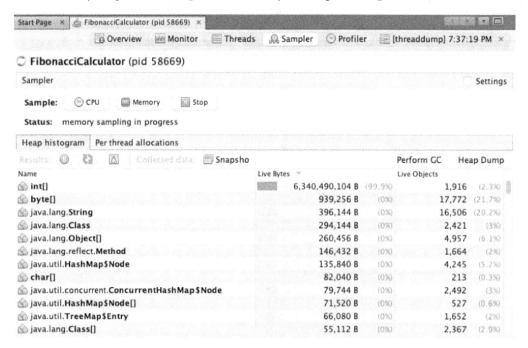

Figure 4.9 – VisualVM – Memory usage using Snapshot

- **Profiler**: This is like the sampler, but it runs all the time. Apart from CPU and memory, we can also look at JDBC invocations and the time it takes to get the response. The next screenshot shows CPU profiling:

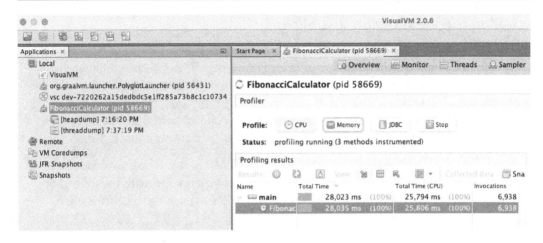

Figure 4.10 – VisualVM – application profiler

Apart from this, Graal VisualVM can be used to analyze Core dumps and to identify the root cause of any Java process crashes. At the time of writing this book, Graal VisualVM supports JavaScript and Ruby (heap, object view, and thread view only), Python, and R (heap and object view only).

JDK Flight Recorder (JFR) analysis is another powerful feature of VisualVM. It helps us to analyze the data that is connected by JFR with no overhead on the running process. JFR provides more advanced analysis, including capturing and analyzing file I/O, socket I/O, and thread locks apart from CPU and thread.

Graal VisualVM also provides extension APIs so we can write custom plugins. Various plugins can be used to extend Graal VisualVM. Here are some of the most widely used plugins:

- **Visual GC plugin**: This plugin provides a powerful interface to monitor garbage collection, class loader, and JIT compiler performance. It is a very powerful plugin that can identify optimizations in the code to improve performance.

- **Tracer**: Tracer provides a better user interface for detailed monitoring and analyzing of the applications.

- **Startup Profiler**: As the name suggests, this provides instrumentation to profile the startups and identify any optimizations that can be performed to improve the startups.

You can find the full list of available plugins at `https://visualvm.github.io/pluginscenters.html`.

Installing the Ideal Graph Visualizer

Ideal Graph Visualizer is a very powerful tool for analyzing how Graal JIT is performing various optimizations. This requires an advanced understanding of Graal Graphs, which is an intermediate representation. Later in this chapter, we will cover Graal Graph and how to use the Ideal Graph Visualizer so that we can see how Graal performs various optimizations. This is critical, as it helps us write better code and optimize the code at development time, and reduces the load on the compiler to perform it just in time.

The The Ideal Graph Visualizer is available with GraalVM EE. It can be downloaded from the Oracle website. The Ideal Graph Visualizer can be launched with the following command, after setting the PATH to the location where it has been unzipped/installed:

```
idealgraphvisualizer
```

The --jdkhome flag can be used to point to the right version of GraalVM. Once it has been launched, you will see the following screen:

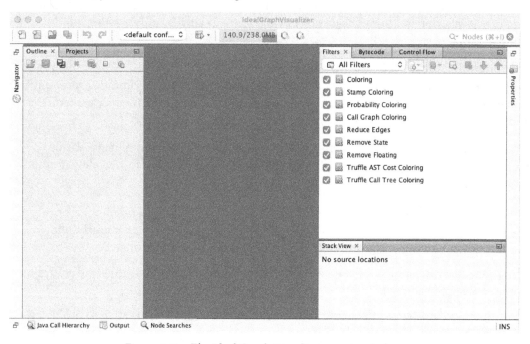

Figure 4.11 – The Ideal Graph Visualizer – main window

The Ideal Graph Visualizer requires Graal dumps to render and analyze Graal Graphs. Graal dumps can be created using the following command:

```
java -Dgraal.Dump=:n <java class file>
```

The n in the preceding command can be 1, 2, or 3, and each number denotes a level of verbosity. This generates a folder called `graal_dumps`, which consists of `bgv` files (Binary Graph Files). Sometimes you will find various `bgv` files due to invalidation and recompilation (deoptimization or on-stack replacements – please refer to *Chapter 2, JIT, HotSpot, and GraalJIT* , and find the *On-stack replacement* section to find out more). These `bgv` files can be opened in The Ideal Graph Visualizer to do the analysis. Once the `bgv` file is loaded, you will see a screen like this:

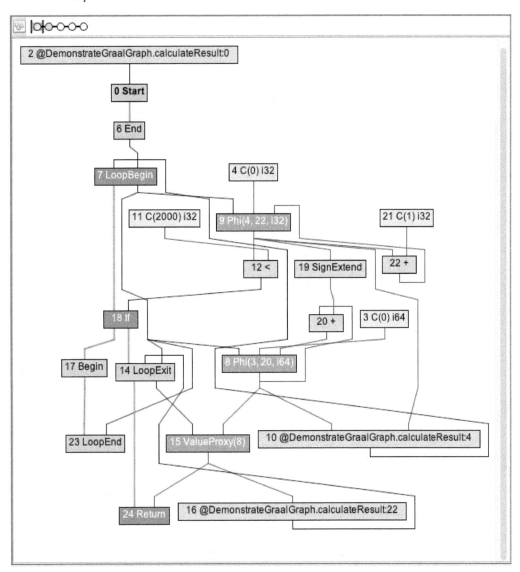

Figure 4.12 – The Ideal Graph Visualizer – main window–Graal dump

The left pane can be used to navigate through the various phases of compilation and optimization, the main window shows the graph, and the right pane can be used to configure how to render these graphs. We can view the Graal Graphs, Call Graph, AST, and Truffle Call Tree.

The Ideal Graph Visualizer can also be connected from the Java runtime (using the `Dgraal.PrintGraph=Network` flag) to view the graphs in real time, while the application code is executing.

In the next section, we will explore how these Graal Graphs can be read to understand how the Graal compiler works.

Understanding the Graal JIT compiler

In the previous chapter, we briefly touched upon the Graal compiler and the ecosystem around it. In this section, we will dig deeper into various compiler options and see how Graal optimizes the code just in time. In the next section, we will take a look at Ahead-of-Time compilation, and how a native image can be created. Before we get into the details of how the Graal compiler works, let's quickly go through some of the Graal compiler configurations, which can be passed as arguments to the virtual machine.

Graal compiler configuration

The Graal compiler can be configured with various arguments that can be passed from the `java` command (in the GraalVM version of `java`). In this section, we will go through some of the most useful command-line configurations.

We will be trying these various flags on a sample application to see how it affects the Graal compiler.

Let's write a simple Java class called `FibonacciCalculator`. Here is the source code of the class:

```java
class FibonacciCalculator{
    public int[] findFibonacci(int count) {
        int fib1 = 0;
        int fib2 = 1;
        int currentFib, index;
        int [] fibNumbersArray = new int[count];
        for(index=2; index < count; ++index ) {
            currentFib = fib1 + fib2;
            fib1 = fib2;
```

```
            fib2 = currentFib;
            fibNumbersArray[index - 1] = currentFib;
        }
        return fibNumbersArray;
    }
    public static void main(String args[])
    {
        FibonacciCalculator fibCal =
            new FibonacciCalculator();
        long startTime = System.currentTimeMillis();
        long now = 0;
        long last = startTime;
        for (int i = 1000000000; i < 1000000010; i++) {
            int[] fibs = fibCal.findFibonacci(i);
            long total = 0;
            for (int j=0; j<fibs.length; j++) {
                total += fibs[j];
            }
            now = System.currentTimeMillis();
            System.out.printf("%d (%d ms)%n", i , now - last);
            last = now;
        }
        long endTime = System.currentTimeMillis();
        System.out.printf ("total: (%d ms)%n",
            System.currentTimeMillis() - startTime);
    }
}
```

As you can see, we are generating `1000000000` to `1000000010` Fibonacci numbers, and then later calculating the sum total of all the Fibonacci number that are generated. The code is written to loops to trigger the compilation threshold.

There are a lot of optimization opportunities for JIT. Let's first run this program with Java HotSpot:

```
↳  chapter4 git:(main) ✗ java -version
java version "15.0.2" 2021-01-19
Java(TM) SE Runtime Environment (build 15.0.2+7-27)
Java HotSpot(TM) 64-Bit Server VM (build 15.0.2+7-27, mixed mode, sharing)
↳  chapter4 git:(main) ✗ java FibonacciCalculator
1000000000 (2626 ms)
1000000001 (1548 ms)
1000000002 (1310 ms)
1000000003 (1307 ms)
1000000004 (1297 ms)
1000000005 (1314 ms)
1000000006 (1324 ms)
1000000007 (1320 ms)
1000000008 (1298 ms)
1000000009 (1323 ms)
total: (14667 ms)
```

Figure 4.13 – FibonnaciCalculator – Java HotSpot output

As you can see, the initial iterations took the most time, and it optimized to around 1,300 ms over the iterations. Let's now compile the code with `javac`, which we got from the Graal EE distribution, and run the same program with Graal JIT. The following screenshot shows the output of running the same application with GraalVM (GraalVM EE 21.0.0.2 on Java 11):

```
→  chapter4 git:(main) ✗ java -version
java version "11.0.10" 2021-01-19 LTS
Java(TM) SE Runtime Environment GraalVM EE 21.0.0.2 (build 11.0.10+8-LTS-jvmci-21.0-b06)
Java HotSpot(TM) 64-Bit Server VM GraalVM EE 21.0.0.2 (build 11.0.10+8-LTS-jvmci-21.0-b06, mixed mode, sharing)
→  chapter4 git:(main) ✗ java FibonacciCalculator
1000000000 (2700 ms)
1000000001 (678 ms)
1000000002 (1136 ms)
1000000003 (973 ms)
1000000004 (852 ms)
1000000005 (857 ms)
1000000006 (823 ms)
1000000007 (826 ms)
1000000008 (822 ms)
1000000009 (824 ms)
total: (10492 ms)
```

Figure 4.14 – FibonnaciCalculator – GraalVM output

We can see significant improvement in performance. Graal started similarly to Java HotSpot, but over the iterations it optimized to 852 ms, compared to the 1,300 ms it took to run with HotSpot. The following option is used to disable GraalJIT and use HotSpot on GraalVM:

```
-XX:-UseJVMCICompiler
```

This is normally used to compare the performance of Graal. Let's run this option with the preceding source code, with the GraalVM EE 21.0.0.2 compiler:

```
java -XX:-UseJVMCICompiler FibonacciCalculator/
```

The following is a screenshot of the output after running the preceding command:

```
→  chapter4 git:(main) ✗ java --version
java 11.0.10 2021-01-19 LTS
Java(TM) SE Runtime Environment GraalVM EE 21.0.0.2 (build 11.0.10+8-LTS-jvmci-21.0-b06)
Java HotSpot(TM) 64-Bit Server VM GraalVM EE 21.0.0.2 (build 11.0.10+8-LTS-jvmci-21.0-b06, mixed mode, sharing)
→  chapter4 git:(main) ✗ java -XX:-UseJVMCICompiler FibonacciCalculator
1000000000 (2178 ms)
1000000001 (1753 ms)
1000000002 (1564 ms)
1000000003 (1525 ms)
1000000004 (1547 ms)
1000000005 (1578 ms)
1000000006 (1638 ms)
1000000007 (1581 ms)
1000000008 (1569 ms)
1000000009 (1527 ms)
total: (16460 ms)
```

Figure 4.15 – FibonnaciCalculator – GraalVM (21/Java 11) output

As you can see, even though we are using the Graal compiler, the performance is similar to Java HotSpot, and in fact is slower than Java HotSpot 15. Note that our Graal is running on Java 11.

The `CompilerConfiguration` flag is used to specify which JIT compiler is to be used. The following is the argument that we can pass to set the compiler configuration:

```
-Dgraal.CompilerConfiguration
```

We have three options; let's also run these options with our sample code to see how it performs:

- `-Dgraal.CompilerConfiguration=enterprise`: This uses the enterprise JIT, and generates the optimum code. However, there will be initial slowdowns due to compilation:

```
→  chapter4 git:(main) ✗ java -Dgraal.CompilerConfiguration=enterprise FibonacciCalculator
1000000000 (2891 ms)
1000000001 (619 ms)
1000000002 (1034 ms)
1000000003 (827 ms)
1000000004 (825 ms)
1000000005 (820 ms)
1000000006 (828 ms)
1000000007 (839 ms)
1000000008 (825 ms)
1000000009 (829 ms)
total: (10337 ms)
```

Figure 4.16 – FibonnaciCalculator – enterprise compiler configuration

- -Dgraal.CompilerConfiguration=community: This produces the community version of JIT, which optimizes to a decent extent. The compilation is therefore faster.

```
→  chapter4 git:(main) x java –Dgraal.CompilerConfiguration=community FibonacciCalculator
1000000000 (2695 ms)
1000000001 (862 ms)
1000000002 (1781 ms)
1000000003 (1567 ms)
1000000004 (1559 ms)
1000000005 (1560 ms)
1000000006 (1558 ms)
1000000007 (1565 ms)
1000000008 (1564 ms)
1000000009 (1568 ms)
total: (16280 ms)
```

Figure 4.17 – FibonnaciCalculator – community compiler configuration

- -Dgraal.CompilerConfiguration=economy: This compiles quickly, with fewer optimizations:

```
→  chapter4 git:(main) x java –Dgraal.CompilerConfiguration=economy FibonacciCalculator
1000000000 (4173 ms)
1000000001 (2541 ms)
1000000002 (2011 ms)
1000000003 (1798 ms)
1000000004 (1883 ms)
1000000005 (1822 ms)
1000000006 (1833 ms)
1000000007 (1831 ms)
1000000008 (1848 ms)
1000000009 (1853 ms)
total: (21594 ms)
```

Figure 4.18 – FibonnaciCalculator – economy compiler configuration

We can see a significant difference in the performance when using enterprise, community, and economy. Here is a comparison of the performances of three options:

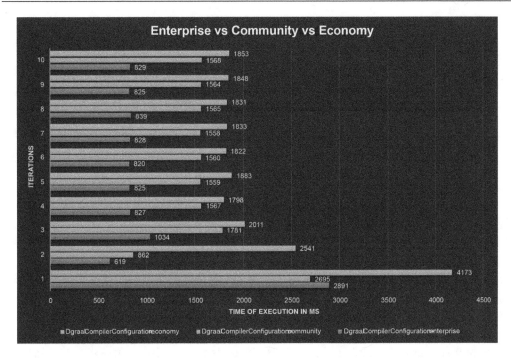

Figure 4.19 – FibonnaciCalculator – enterprise versus community versus economy configuration

Apart from this, there are a lot of other performance tuning options that can be used to improve the performance of the compiler, such as this:

```
-Dgraal.UsePriorityInlining (true/false)
```

The preceding flag can be used to enable/disable advanced inlining algorithm. Disabling this improves the compilation time and helps throughput.

This flag can be used to disable auto-vectorization optimization:

```
-Dgraal.Vectorization (true/false)
```

This flag can be used to disable the path duplication optimization, such as Dominance-Based Duplication Simulation. When this is disabled, it has an impact on the throughput:

```
-Dgraal.OptDuplication (true/false)
```
```
This next flag can be set to values between -1 and 1. When the
value is below 0, the JIT reduces the effort spent on inlining.
This will improve the startup and provides throughput. When
the value is greater than 0, the JIT spends more effort in
inlining, increasing the performance:
```
```
-Dgraal.TuneInlinerExploration (-1 to +1)
```

This is a very useful flag that can be enabled to trace how the JIT compiler takes decisions on inlining optimization:

```
-Dgraal.TraceInlining (true/false)
```

When we enable this flag for sample code, we get the following:

```
compilation of FibonacciCalculator.main(String[]):
  at FibonacciCalculator.main(FibonacciCalculator.java:20)
[bci: 4]: <GraphBuilderPhase> FibonacciCalculator.<init>():
yes, inline method
  at FibonacciCalculator.main(FibonacciCalculator.java:25)
[bci: 32]: <GraphBuilderPhase> FibonacciCalculator.
findFibonacci(int): no, bytecode parser did not replace invoke
```

```
compilation of FibonacciCalculator.main(String[]):
  at FibonacciCalculator.main(FibonacciCalculator.java:20)
[bci: 4]: <GraphBuilderPhase> FibonacciCalculator.<init>():
yes, inline method
  at FibonacciCalculator.main(FibonacciCalculator.java:25)
[bci: 32]:
    ├──<GraphBuilderPhase> FibonacciCalculator.
findFibonacci(int): no, bytecode parser did not replace invoke
    └──<PriorityInliningPhase> FibonacciCalculator.
findFibonacci(int): yes, worth inlining according to the cost-
benefit analysis.
```

```
compilation of java.lang.String.hashCode():
  at java.lang.String.hashCode(String.java:1504) [bci: 19]:
    ├──<GraphBuilderPhase> java.lang.String.isLatin1(): no,
bytecode parser did not replace invoke
    └──<PriorityInliningPhase> java.lang.String.isLatin1(): yes,
budget was large enough to inline this callsite.
  at java.lang.String.hashCode(String.java:1504) [bci: 29]:
    ├──<GraphBuilderPhase> java.lang.StringLatin1.
hashCode(byte[]): no, bytecode parser did not replace invoke
    └──<PriorityInliningPhase> java.lang.StringLatin1.
hashCode(byte[]): yes, budget was large enough to inline this
callsite.
```

We can see how the JIT compiler is taking decisions on inlining.

These optimization flags can be set even for other GraalVM launchers, such as `js` (for JavaScript), `node`, and `lli`.

Graal JIT compilation pipeline and tiered optimization

In the previous chapter, in the *Graal JIT compiler* section, we looked at how Graal JIT integrates with the virtual machine through JVMCI. In this section, let's take a deeper look at how Graal JIT interacts with virtual machine.

Graal optimizes the code in three tiers. The tiered approach helps Graal to perform optimizations starting from more platform-independent representations (high-level intermediate representation) to more platform-dependent representations (low-level intermediate representation). The following diagram shows how Graal JIT interfaces with the virtual machine and performs these three tiers of optimization:

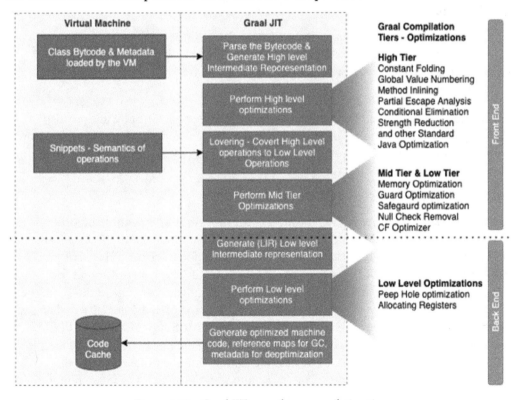

Figure 4.20 – Graal JIT compiler – compilation tiers

Let's try to understand this picture better:

- The virtual machine passes the bytecode and metadata to the Graal JIT when it hits the compilation threshold (refer to *Chapter 2, JIT, HotSpot, and GraalJIT*, to find out more about the compilation thresholds).

- Graal parses the bytecode and generates a **high-level intermediate representation (HIR)**.

- It then performs various optimizations on the **HIR**. These are some of the standard Java optimization techniques that are applied, with some new techniques that have been introduced in Graal, such as partial escape analysis and advanced inlining techniques.

- Once these high-level optimizations are performed, Graal starts converting the high-level operations to low-level operations. This phase is called lowering. There are two tiers of optimizations that it performs during this phase, and it eventually generates the **low-level intermediate representation (LIR)** for the target processor architecture.

- Once all the optimizations are performed on the LIR, the final optimized machine code is generated and stored in the code cache, along with the reference maps that the garbage collector will use and the metadata that will be required for deoptimization.

In this section, we looked at how the Graal JIT compiler works internally, and we also explored various compiler configurations that will affect the compiler's performance. Now let's understand the Graal Intermediate Representation better.

Graal intermediate representation

Intermediate Representations (IRs) are among the most important data structures for compiler design. IRs provide a graph that helps the compiler understand the structure of the code, identify opportunities, and perform optimizations. Selecting the right type of data structure and IR is key for a compiler to identify these opportunities for optimization. In GraalVM, apart from the C1 and C2 compilers that exist in the JVM, the Graal compiler was introduced with a new IR, which is based on a directed graph data structure. Each node in the graph represents a value in **Static Single Assignment (SSA)** form. Since the Graal compiler is completely built on Java, the node types are declarative as class definitions, each node has a Java class definition, and the node types are all defined as a hierarchy of Java classes. The operations and the values, which are represented as nodes, are defined by their corresponding types, for example `AddNode`, `IfNode`, and `SwitchNode`, all of them deriving from the base class, `Node`. The edges (operands) are represented as fields of the class. The following diagram shows the hierarchy of various types of node:

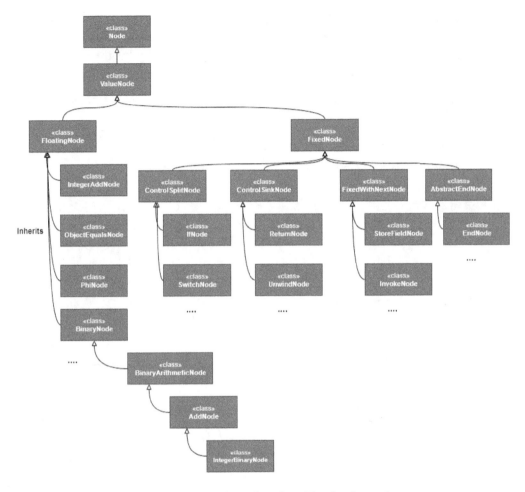

Figure 4.21 – Graal Graph nodes – The class hierarchy

The representation of code in SSA enables the creation of a single version of the variable for each value. This helps perform better data flow analysis and optimizations. A phi function (Φ) is used to convert decision-based control paths (such as `if` and `switch`). The Phi function is a function of two values, and the value is selected based on the control flow. Refer to the following papers on SSA for more details: `https://gcc.gnu.org/onlinedocs/gccint/SSA.html` and `https://en.wikipedia.org/wiki/Static_single_assignment_form`. The key point is that the complete program is converted into an SSA to perform the optimizations.

Graal IRs are built as a Graal graph, where each node has input edges that point to the nodes that create the operands and the successor edges that show the control flow. The successor edge points to the node that succeeds the current node in terms of the control flow.

To demonstrate everything that we have discussed so far, let's analyze some simple Java code with The Ideal Graph Visualizer. The logic in this code may not generate a simple graph – the code is intentionally kept simple. The loops are there to hit the threshold so that when the JVM hits the threshold, it will perform Graal JIT compilation, as shown next:

```
public class DemonstrateGraalGraph {

    public long calculateResult() {
        long result = 0;
        for (int i=0; i<2000; i++) {
            result = result + i;
        }
        return result;
    }

    public static void main(String[] args) {
        DemonstrateGraalGraph obj =
            new DemonstrateGraalGraph();
        while (true) {
//This loop is just to reach the compiler threshold
            long result = obj.calculateResult();
            System.out.println("Total: " + result);
        }
    }
}
```

Let's now compile the preceding code using the `javac DemonstrateGraalGraph.java` command. To keep the graph simple, we will compile only the `calculateResult()` method by using the `-XX:CompileOnly=DemonstrateGraalGraph:calculateResult` flag. Let's also disable some optimizations using the following flags:

`-Dgraal.FullUnroll=false`, `-Dgraal.PartialUnroll=false`, `-Dgraal.LoopPeeling=false`, `-Dgraal.LoopUnswitch=false`, `-Dgraal.OptScheduleOutOfLoops=false`, and `-Dgraal.VectorizeLoops=false`

So, we get the following:

```
java -XX:CompileOnly=DemonstrateGraalGraph::calculateResult \
 -XX:-UseOnStackReplacement \
 -Dgraal.Dump=:1 \
 -XX:+PrintCompilation \
  -Dgraal.FullUnroll=false \
  -Dgraal.PartialUnroll=false \
  -Dgraal.LoopPeeling=false \
  -Dgraal.LoopUnswitch=false \
  -Dgraal.OptScheduleOutOfLoops=false \
  -Dgraal.VectorizeLoops=false   \
 DemonstrateGraalGraph
```

This will create a folder called graal_dumps with a dump of all the Graal JIT activities. Once you load the bgv file that is generated by Graal, you will find the various tiers of optimization listed in the left pane, as shown in the following screenshot:

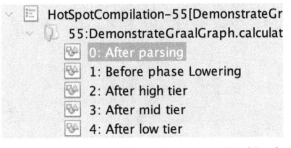

Figure 4.22 – The Ideal Graph Visualizer – DemonstrateGraalGraph – left pane

When you click on **0: After parsing** on the right page, you will see the Graal graph representation after parsing the bytecode. In the graph, the red lines represent the control flow and the blue lines represent the data flow. The control flow has to be read from top to bottom, but data flow normally can be understood by reading upward. Let's understand this picture and compare it with the code. Please note that this graph is only for the `calculateResult()` method as we asked the JVM to create only compile `calculateResult()` method. Let's understand this graph better:

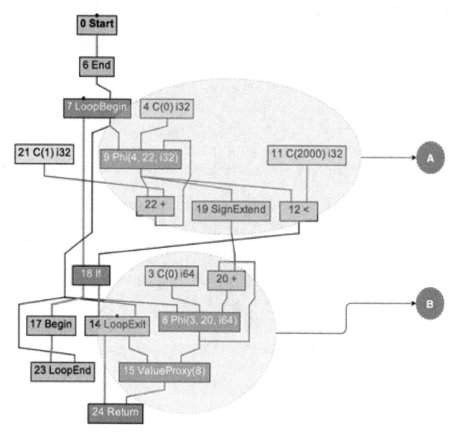

Figure 4.23 – The Ideal Graph Visualizer – DemonstrateGraalGraph – Graal Graph after parsing

The program starts with **0 Start** and the loop starts at the **7 LoopBegin** node. To make the graph easier to understand, some of the sections are highlighted with the labels **A** and **B**. Let's explore what these sections of the graph are.

Section A

- Section A highlights the `for` loop. It is converted into a `18  if` statement. The input for the `if` statement is the current value of I, which is the output of the Phi node 9 `Phi(4,22,i32)` and constant 2000 node 11 `C(2000)  i32`.

- Phi is attached where the control flows merge. In this case, *9 Phi (4,22, i32)* merges the output from 4 `C(0)  i32` (`i=0` in the `for` loop) and the output of the 22 + node (which is `i++`). This node will simply output the current value of the `i` after incrementing by the value of the **21 C(1) i32 node**.

- This then flows into the **12 < node** and is compared with **11 C(2000) i32** (which is the maximum value of the loop), and this expression is evaluated by control flow node **18 if**.

Section B

- Section B highlights the section where the result is calculated.

- The initial value of the result is represented as **C (0) i64**. It is `i64`, as we declared it as a `long`.

- The **8 Phi(3, 20, i64)** node merges the control flow to calculate the `result  = result  +  i` expression. The value of `i` is flowing from the **19 SignExtend** node, which is an output of the current value of I, which flows from **9 Phi(4,22, i32)**.

- The final out flows into **24 Return** when the loop ends at **18 if**.

Now we can go through each phase of optimization by selecting the phase in the left pane to see how the code is optimized. Let's quickly look at how this graph is transformed through the phases. When we select **Before Phase Lowering** in the left pane's **Outline** window, we will see the following graph:

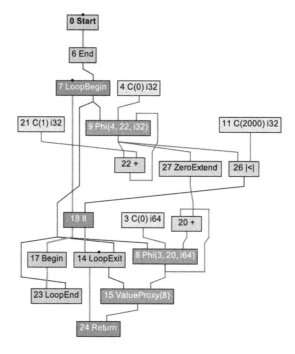

Figure 4.24 – The Ideal Graph Visualizer – DemonstrateGraalGraph – Graal graph before lowering

In this phase, we can see the following optimizations:

- The **19 Sign Extend** node is replaced with **27 Zero Extend**, as the compiler found out that it is an unsigned integer. Operations with unsigned integers are less expensive than operations with signed integers.

- The **12 <** node is replaced with **26 |<|**, which is an unsigned less than operation, which is faster. The compiler arrives at this conclusion based on the various iterations and profiling. Since the operands are considered unsigned, even the operations are considered unsigned.

- The graph also illustrates application of the canonicalization technique of replacing <= with <, to speed up the if (which is originally the for loop) statements.

The subsequent phases – high tier, mid tier, and low tier – may not show significant optimizations, as the code is relatively simple and we have disabled some of the optimizations to keep the graph simple to read and understand:

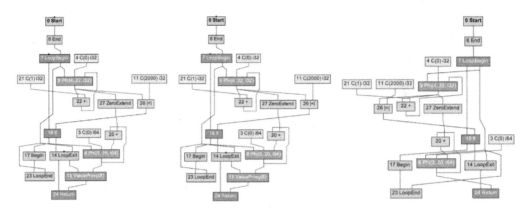

Figure 4.25 – The Ideal Graph Visualizer – DemonstrateGraalGraph – Graal Graph, other tiers

Figure 4.26 is a diagram of the graph with all optimizations enabled. You will see that loop unrolling has been used very prominently to speed up the loops:

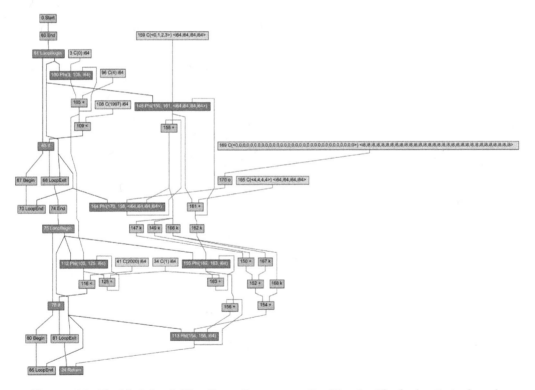

Figure 4.26 – The Ideal Graph Visualizer – DemonstrateGraalGraph – The final optimized graph

Graal performs various optimizations as part of the tiered compilation. We will go through this in detail in the next section, and also see how we can use this knowledge to improve the way we write the code.

Understanding Graal compiler optimizations

The Graal compiler performs some of the most advanced optimizations on the code just in time. The most critical ones are discussed in the following subsections.

Before getting into this session, please refer to the *Understanding the optimizations performed by JIT* section of *Chapter 2, JIT, HotSpot, and GraalJIT*.

Speculative optimization

JIT compilation relies heavily on the runtime profiling of the code. As we have seen, the graphs are optimized based on the HotSpots. HotSpots, as we covered in *Chapter 2, JIT, HotSpot, and GraalJIT*, are the control flows that the program goes through most frequently. There is no point in trying to optimize the whole code; instead, the JIT compiler tries to optimize the hot control paths/flows. This is based on speculation and assumption. When an assumption is proven wrong during execution, the compiler quickly deoptimizes and waits for another opportunity to optimize based on new HotSpots. We covered compiler thresholds and Hot Spots in *Chapter 2, JIT, HotSpot, and GraalJIT*, in the *Compiler threshold* section. Graal JIT also uses similar techniques to identify the hotspots. Graal performs all the optimizations that we discussed in *Chapter 2, JIT, Hotspot, and GraalJIT*, in the *Understanding the optimizations performed by JIT* section, but also uses some advanced techniques. Let's go through some of the most important optimization techniques that Graal JIT applies to the code.

Partial escape analysis

In *Chapter 2, JIT, HotSpot, and GraalJIT*, in the section titled *Understanding the optimizations performed by JIT*, we explored escape analysis. Escape analysis is one of the most powerful techniques. It identifies the object's scope and the objects escape from local to global scope. If it identifies objects that don't escape, there is an opportunity to optimize, and the compiler will optimize the code to use stack allocation instead of heap allocation for the objects that are within the local scope. This saves a significant amount of time in allocating and deallocating memory in the heap.

Partial escape analysis takes this further by not just limiting to identifying objects that escape a method level scope to control branches. This helps to optimize the code when an object is found to be escaping only in certain control flows. Other control flows where the object is not escaping can be optimized to either use a local value or scalar replacements.

Partial escape analysis looks for escapes that might happen through method calls, return values, throw statements, and so on. Let's use a simple code to understand how this works:

```
public void method(boolean flag) {
    Class1 object1 = new Class1();

    Class2 object2 = new Class2();
    //some processing
    object1.parameter = value;
    //some more logic
    if(flag) {
        return object1;
    }
    return object2;
}
```

The preceding is some sample code, just to illustrate partial escape analysis. In this code we are creating object1 as an instance of Class1 and object2 as an instance of Class2. Some processing is happening, and object1 fields are updated with some values that are calculated. Based on the flag, object1 or object2 will escape. Let's assume that most of the time the flag is false, and only object1 escapes, so there is no point in creating object1 every time the method is called. This code gets optimized to something like the following (this is just an illustration of how partial escape analysis works; the Graal JIT may not do this exact refactoring):

```
public void method(boolean flag) {

    Class2 object2 = new Class2();

    tempValue = value;

    if(flag) {
        Class1 object1 = new Class1();
        object1.parameter = tempValue;
```

```
        return object1;
    }
    return object2;
}
```

`object1` is only created if required, and a temporary variable is used to store the intermediate values, and if `object1` has to be initialized then it uses the temporary values before it escapes. This optimizes heap allocation time and heap size.

Inter-procedural analysis and inlining

Graal performs optimization at the AST/Graph level. This helps Graal to perform inter-procedural analysis and identify any option that may never become empty and skips compiling that part of the code, as it may never be called. It adds a guard to that code block, just in case. If the control flows through that block, the JIT can deoptimize the code.

To understand inter-procedural analysis and inlining, one commonly used example is a JDK class, `OptionalDouble`. Here is a snippet of the `OptionalDouble` class:

```
public class OptionalDouble {
    public double getAsDouble() {
        if (!isPresent) {
            throw new
                NoSuchElementException("No valuepresent");
        }
        return value;
    }
}
```

Let's say we call this `getAsDouble()` method, and the method has a `throw` block, but that `throw` block may never be called. The Graal compiler will compile all the code, except the `if` block, and will place a `guard` statement so that, if it gets called, it can deoptimize the code. Apart from this, Graal performs more advanced inlining to optimize the code. We can look at the complete set of optimizations that Graal performs by passing `-Dgraal.Dump=:2`. With Graal dump at level 2, we get a much more detailed list of graphs for each phase. In the next screenshot, you can see a whole list of optimizations the Graal JIT performed on the code across the various tiers of compilation:

Figure 4.27 – The Ideal Graph Visualizer – DemonstrateGraalGraph – compilation tiers

By looking at how the graph is optimized at each step, we can see every area where the code can be optimized at development time. This will reduce the load on Graal JIT and the code will perform better. Some of these optimization techniques are covered in the *Understanding the optimizations performed by JIT* section of *Chapter 2, JIT, HotSpot, and GraalJIT*.

Debugging and monitoring applications

GraalVM comes with a rich set of tools for debugging and monitoring applications. We have already looked at VisualVM and the Ideal Graph Visualizer. As you saw in the previous sections, these two tools are very powerful for detailed analysis. This analysis also provides insights into how we can improve the code at the development time to reduce the load on Graal JIT, and write high-performing and low-footprint Java code. Apart from these two tools, the following are some of the other tools that Graal comes with.

Seafoam

Seafoam is a great alternative for Ideal Graph Visualizer. The following are some of the enahnced capabilities that Seafoam provides over IGV:

- Seafoam is a Open source project, and can be used according to MIT license.

- It provides various export options of the graphs, which is a great convince over IGV.

- It provides a CLI, which helps in running it as part of any automation.

- Seafoam also be used asa library, to embed it as part of the application.

- The graphs generated by Seafoam are simpler and easier to understand.

For more latest information on Seafoam, please refer to the GitHub page (`https://github.com/Shopify/seafoam`)

Visual Studio Code extension

The Visual Studio Code extension is one of the most powerful integrated development environments for Graal. The following screenshot shows the GraalVM Extension for Visual Studio Code:

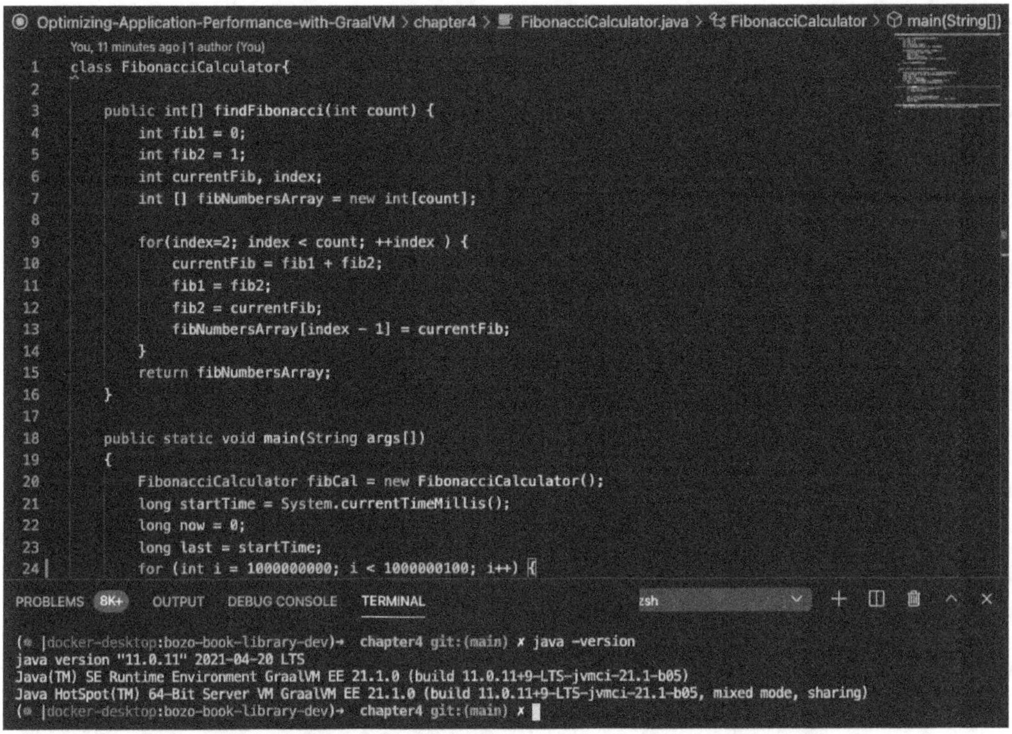

```
Optimizing-Application-Performance-with-GraalVM > chapter4 > 🔖 FibonacciCalculator.java > 🔁 FibonacciCalculator > ⓞ main(String[])
        You, 11 minutes ago | 1 author (You)
 1      class FibonacciCalculator{
 2
 3          public int[] findFibonacci(int count) {
 4              int fib1 = 0;
 5              int fib2 = 1;
 6              int currentFib, index;
 7              int [] fibNumbersArray = new int[count];
 8
 9              for(index=2; index < count; ++index ) {
10                  currentFib = fib1 + fib2;
11                  fib1 = fib2;
12                  fib2 = currentFib;
13                  fibNumbersArray[index - 1] = currentFib;
14              }
15              return fibNumbersArray;
16          }
17
18          public static void main(String args[])
19          {
20              FibonacciCalculator fibCal = new FibonacciCalculator();
21              long startTime = System.currentTimeMillis();
22              long now = 0;
23              long last = startTime;
24              for (int i = 1000000000; i < 1000000100; i++) {
```

```
PROBLEMS  8K+    OUTPUT    DEBUG CONSOLE    TERMINAL                        zsh                ∨   +  ⧉  🗑  ∧  ×

(● |docker-desktop:bozo-book-library-dev)→  chapter4 git:(main) ✗ java —version
java version "11.0.11" 2021-04-20 LTS
Java(TM) SE Runtime Environment GraalVM EE 21.1.0 (build 11.0.11+9-LTS-jvmci-21.1-b05)
Java HotSpot(TM) 64-Bit Server VM GraalVM EE 21.1.0 (build 11.0.11+9-LTS-jvmci-21.1-b05, mixed mode, sharing)
(● |docker-desktop:bozo-book-library-dev)→  chapter4 git:(main) ✗ ▊
```

Figure 4.28 – GraalVM environments on Visual Studio Code

In the previous screenshot, you can see all the various GraalVM installations that have been configured on the left pane. It is very easy to switch between various versions of GraalVM, and the terminal and the whole environment will use the selected GraalVM.

This extension also makes it easy to install optional components. We don't have to manually run gu commands. This extension provides easy way to build, debug, and run code written in Java, Python, R, Ruby, and Polyglot (mixed language code).

This extension can be directly installed from the Visual Studio Code Extensions tab by searching for Graal. The following screenshot shows the extension installation page:

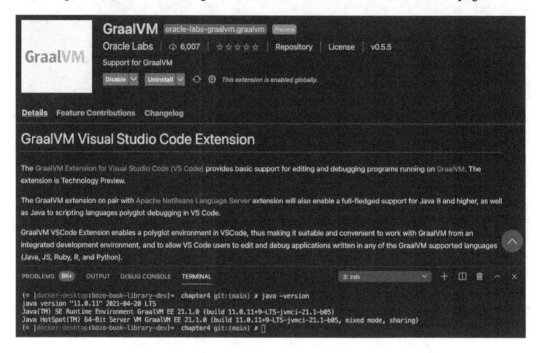

Figure 4.29 – Installing the GraalVM Extension for Visual Studio Code

There is also a GraalVM Extension Pack that comes with additional features such as Micronaut framework integration and NetBeans Language Server, which provide Java code completion, refactoring, Javadoc integration, and many more advanced features. The next screenshot shows the installation page for the GraalVM Extension Pack:

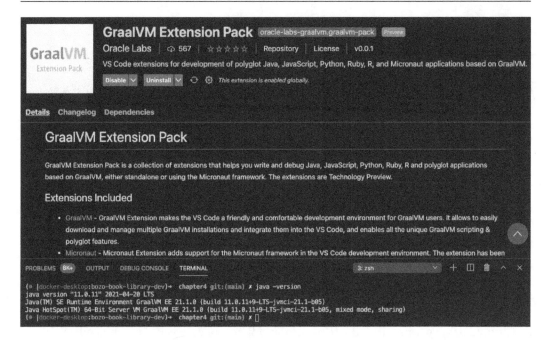

Figure 4.30 – GraalVM Extension Pack plugin for Visual Studio Code

You can learn more about this extension on the GraalVM website at `https://www.graalvm.org/tools/vscode/graalvm-extension/`.

GraalVM Dashboard

GraalVM Dashboard is a web-based tool with which we can perform detailed analysis of both static and dynamic compilations. This is very powerful for Native Image analysis. The tool provides details dashboard reports on compilation, reachability, usability, profiling data, dynamic compilation parameters, deoptimization, and more.

We will run this tool in the next chapter, when we create a native image of our sample code and perform more detailed analysis of the native image code.

Command-line tools

There are two command-line tools that can be used in the context of Polyglot to identify further opportunities the optimize the code. We will be using these tools in *Chapter 6*, *Truffle for Multi-language (Polyglot) support*, for Polyglot optimization. The following are the two command-line tools that come with GraalVM:

- **Profiling CLI**: This tool helps to identify opportunities to optimize CPU and memory usage. Please refer to `https://www.graalvm.org/tools/profiling/` for more details.

- **Code Coverage CLI**: This tools records and analyzes the code coverage for each execution. This is very powerful for running test cases and ensuring good code coverage. This tool can also identify possible dead code that can be eliminated, or hot code that can be optimized at development time. Please refer to `https://www.graalvm.org/tools/code-coverage/` for more details.

Chrome debugger

The Chrome debugger provides the Chrome developer tool extension to debug guest language applications. The Chrome debugger can be used while running the application with the `--inspect` option. This helps debug JavaScript (Node.js) using Chrome. The extension can be installed from `https://developers.google.com/web/tools/chrome-devtools/`.We will be covering this tool while we discuss JavaScript and Node.js on Graal in *Chapter 6*, *Truffle for Multi-language (Polyglot) support*.

Summary

In this chapter, we went through Graal JIT and Ahead of Time compilers in detail. We took a sample code and looked at how Graal JIT performs various optimizations using the Ideal Graph Visualizer. We also went through Graal Graphs in detail. This is very critical knowledge that will help you analyze and identify optimizations that can be applied during development to speed up Graal JIT compilation at runtime.

In this chapter, you have gained a detailed understanding of the internal workings of Graal JIT compilation, and how to fine-tune Graal JIT. You also have gained a good understanding of how to use some of the advanced analysis and diagnosis tools to debug Graal JIT compilation, and identify opportunities to optimize the code.

In the next chapter, we will take a more detailed look at Graal Ahead of Time compilation.

Questions

1. What are the various tiers of Graal JIT compilation?

2. What is an intermediate representation?

3. What is SSA?

4. What is speculative optimization?

5. What is the difference between escape analysis and partial escape analysis?

Further reading

- Partial Escape Analysis and Scalar Replacement for Java (`https://ssw.jku.at/Research/Papers/Stadler14/Stadler2014-CGO-PEA.pdf`)

- Understanding Basic Graal Graphs (`https://chrisseaton.com/truffleruby/basic-graal-graphs/`)

- Optimizing Strategies of GraalVM (`https://www.beyondjava.net/graalvm-plugin-replacement-to-jvm`)

- GraalVM Enterprise Edition (EE) (`https://docs.oracle.com/en/graalvm/enterprise/19/index.html`)

- GraalVM documentation (`https://www.graalvm.org/docs/introduction/`)

- Static Single Assignment (`https://gcc.gnu.org/onlinedocs/gccint/SSA.html`)

5
Graal Ahead-of-Time Compiler and Native Image

Graal ahead-of-time compilation helps build native images that start up faster, and have a smaller footprint than traditional Java applications. Native images are critical for modern-day cloud-native deployments. GraalVM comes bundled with a tool called `native-image` that is used to compile ahead of time and generate native images.

`native-image` compiles the code into a native executable/native binary that can run standalone without a virtual machine. The executable includes all the classes, dependencies, libraries, and more importantly, all the virtual machine functionalities such as memory management, thread management, and so on. The virtual machine functionality is packaged as a runtime called Substrate VM. We briefly covered Substrate VM in *Chapter 3*, *GraalVM Architecture*, in the *Substrate VM (Graal AOT and native image)* section. In this chapter, we will gain a deeper understanding of native images. We will learn how to build, run, and optimize native images with a sample.

Native images can only perform static code optimizations and do not have the advantage of runtime optimizations that just-in-time compilers perform. We will explore profile-guided optimization, which is a technique that can be used to optimize native images, by using runtime profiling data.

In this chapter, we will cover the following topics:

- Understanding how to build and run native images
- Understanding the architecture of a native image, and how the compilation works
- Exploring various tools, compilers, and runtime configurations to analyze and optimize the way native images are built and executed
- Understanding how to optimize native images using **Profile-Guided Optimization (PGO)**
- Understanding the limitations of native images, and how to overcome these limitations
- Understanding how memory is managed by native images

By the end of this chapter, you will have a clear understanding of Graal ahead-of-time compilation and hands-on experience in building and optimizing native images.

Technical requirements

We will be using the following tools and sample code for exploring and understanding Graal ahead-of-time compilation:

- **The native-image tool**: We will cover how to install and run the `native-image` tool.
- **GraalVM Dashboard**: We will be using GraalVM Dashboard in this chapter to analyze the native images that we create.
- **Access to GitHub**: There are some sample code snippets, which are available in a Git repository. The code can be downloaded from `https://github.com/PacktPublishing/Supercharge-Your-Applications-with-GraalVM/tree/main/Chapter05`.
- The Code in Action video for this chapter can be found at `https://bit.ly/3ftfzNr`.

Building native images

In this section, we will build a native image, using the Graal Native Image builder (`native-image`). Let's start by installing the Native Image builder.

`native-image` can be installed using GraalVM Updater with the following command:

```
gu install native-image
```

The tool is directly installed in the `/bin` folder of `GRAALVM_HOME`.

Let's now create a native image of `FibonacciCalculator`, from the *Graal compiler configurations* section in *Chapter 4, Graal Just-In-Time Compiler.*

To create a native image, compile the Java file and run `native-image FibonacciCalculator -no-fallback -noserver`. The following screenshot shows the output after running the command:

```
→ chapter4 git:(main) ✗ native-image FibonacciCalculator
[fibonaccicalculator:19526]       classlist:      921.60 ms,   0.93 GB
[fibonaccicalculator:19526]           (cap):    6,533.45 ms,   0.93 GB
[fibonaccicalculator:19526]         setup:    14,575.11 ms,   0.93 GB
[fibonaccicalculator:19526]        (clinit):      167.10 ms,   1.19 GB
[fibonaccicalculator:19526]      (typeflow):    3,647.66 ms,   1.19 GB
[fibonaccicalculator:19526]       (objects):    3,284.82 ms,   1.19 GB
[fibonaccicalculator:19526]      (features):      186.39 ms,   1.19 GB
[fibonaccicalculator:19526]        analysis:    7,436.04 ms,   1.19 GB
[fibonaccicalculator:19526]        universe:      341.78 ms,   1.19 GB
[fibonaccicalculator:19526]         (parse):      793.61 ms,   1.49 GB
[fibonaccicalculator:19526]        (inline):    1,158.16 ms,   1.49 GB
[fibonaccicalculator:19526]       (compile):   11,030.57 ms,   3.90 GB
[fibonaccicalculator:19526]         compile:   13,532.39 ms,   3.90 GB
[fibonaccicalculator:19526]           image:    1,160.83 ms,   3.90 GB
[fibonaccicalculator:19526]           write:    2,669.70 ms,   3.90 GB
[fibonaccicalculator:19526]         [total]:   40,813.57 ms,   3.90 GB
```

Figure 5.1 – FibonacciCalculator – generating a Native Image console output

Native Image compilation takes time, as it has to perform a lot of static code analysis to optimize the image that is generated. The following diagram shows the flow of ahead-of-time compilation performed by the Native Image builder:

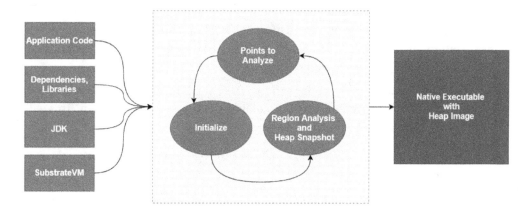

Figure 5.2 – Native Image pipeline flow

Let's try to understand this picture better:

- The ahead-of-time compiler loads all the application code and dependency libraries and classes and packages them along with the Java Development Kit classes and Substrate VM classes.

- Substrate VM has all the virtual machine functionality that is required to run the application. This includes memory management, garbage collection, thread management, scheduling, and so on.

- The compiler then performs the following optimization on the code before building the native image:

  a. **Points-to Analysis**: Identifies only classes and methods that are used and called, and eliminates all the code that is never used or called. For example, our code may not be doing a lot of math operations or string operations. There is no point in taking the complete JDK and converting it to machine code. Instead, only the used classes/methods are picked. The points-to analysis assumes that all the bytecode that is ever used in the application is available. This means that the classes cannot be loaded at runtime. Even the runtime modification of class structure, using reflection, is not supported directly. However, if the application uses reflection, we can create a meta-file in JSON, to configure reflection (META-INF/native-image/reflect-config.json). It is also possible to configure other dynamic features such as JNI, proxies, and so on. The configuration files need to be in CLASSPATH, then the compiler takes care of including these features in the final native image.

  b. **Initialize**: The classes are initialized and the heap allocations are performed.

  c. **Region Analysis and Heap Snapshot**: Create a snapshot of the heap so that the heap image can be created and built into the native image, for faster loads. Heap snapshotting gathers all the objects that are reachable at runtime. This includes class initializations, initializing static and static final fields, enum constants, java.lang.Class objects, and so on.

- The final native image has the code section, where the final optimized binary code is placed, and in the data section of the executable, the heap image is written. This will help load the native image quickly.

Here is a higher-level flow of how points-to analysis and region analysis work at build time and runtime:

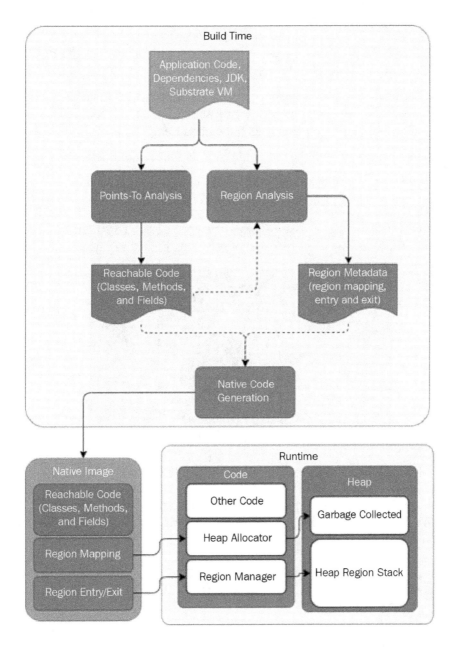

Figure 5.3 – Native Image pipeline – points-to analysis and region analysis

Let's understand this picture in detail:

- At build time, the points-to analysis scans through the application code, dependencies, and JDK to find reachable code.

- The region analysis captures the heap region metadata, which includes region mappings and the region entry/exit. Region analysis also uses the reachable code to identify which static elements need to be initialized.

- The code is generated with the reachable code and the region metadata.

- At runtime, the heap allocator allocates ahead of time using the region mappings and the region manager handles the entry and exit.

Let's now run the native image by issuing `./fibonaccicalculator` from the command line. The following is a screenshot of executing the native image:

```
→  chapter4 git:(main) x ./fibonaccicalculator
1000000000 (3029 ms)
1000000001 (3096 ms)
1000000002 (3116 ms)
1000000003 (3089 ms)
1000000004 (3170 ms)
1000000005 (3102 ms)
1000000006 (3124 ms)
1000000007 (3063 ms)
1000000008 (3050 ms)
1000000009 (3068 ms)
total: (30907 ms)
```

Figure 5.4 – FibonacciCalculator – running fibonaccicalculator native image console output

One of the biggest disadvantages of ahead-of-time compilation is that the compiler never gets to profile the runtime to optimize the code, which happens very well with just-in-time compilation. To bring the best of both worlds together, we can use the PGO technique. We covered PGO briefly in *Chapter 3, GraalVM Architecture*. Let's see it in action and understand it in a little more depth.

Analyzing the native image with GraalVM Dashboard

To gain a deeper understanding of how the points-to analysis and region analysis works, we can use GraalVM Dashboard. In this section, we will create a dump while building the native image and use GraalVM to visualize the Native Image builder perform points-to analysis and region analysis.

In the section *Debugging and Monitoring applications* from *Chapter 4, Graal Just-In-Time Compiler,* we briefly covered GraalVM Dashboard. GraalVM Dashboard is a very powerful tool specifically for native images. In this section, we will generate a dashboard dump of our `FibonnacciCalculator` sample, and explore how we can use GraalVM Dashboard to gain insights into the native image.

To generate the dashboard dump, we have to use the `-H:DashboardDump=<name of the file>` flag. For our `FibonacciCalculator`, we use the following command:

```
native-image -H:DashboardDump=dashboard -H:DashboardAll
FibonacciCalculator
```

The following screenshot shows the output generated by this command. The command created a `dashboard.bgv` file:

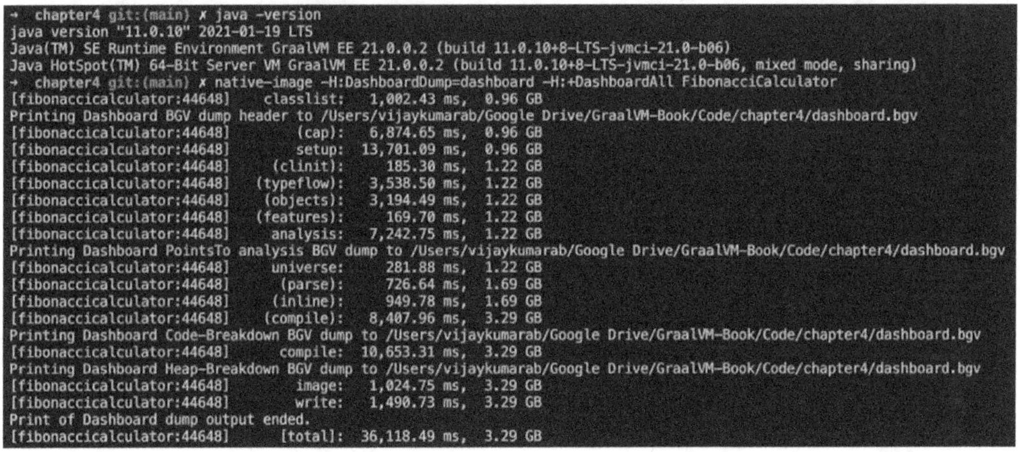

Figure 5.5 – FibonacciCalculator – generating the dashboard dump console output

We also used the `-H:DashboardAll` flag to dump all the parameters. The following are alternative flags that we can use:

- `-H:+DashboardHeap`: This flag only dumps the image heap.

- `-H:+DashboardCode`: This flag generates the code size, broken down by method.

- `-H:+DashboardPointsTo`: This flag creates a dump of the points-to analysis that the Native Image builder has performed.

Now let's load this `dashboard.bgv`, and analyze the results. We need to upload the `dashboard.bgv` file to GraalVM Dashboard. Open the browser and go to `https://www.graalvm.org/docs/tools/dashboard/`.

We should then see the following screen:

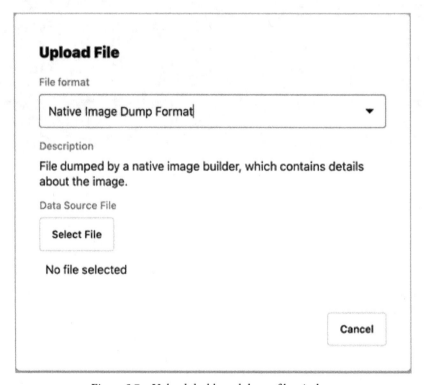

Figure 5.6 – GraalVM Dashboard home page

Click on the **+ Load data** button at the top left. You will get a dialog box as shown in the next screenshot:

Upload File

File format

Native Image Dump Format

Description

File dumped by a native image builder, which contains details about the image.

Data Source File

Select File

No file selected

Cancel

Figure 5.7 – Upload dashboard dump file window

Click on the **Select File** button and point to the `dashboard.bgv` file that we generated. You will immediately see the dashboard, as shown in *Figure 5.8*. You will find two reports that are generated on the left side – **Code Size Breakdown** and **Heap Size Breakdown**.

Understanding the code size breakdown report

The code size breakdown report provides the size of the code of various classes categorized into blocks. The size of the block represents the size of the code. The following figure shows the initial dashboard screen when we select the **Code Size Breakdown** option in the left pane for `dashboard.bgv` we generated in the previous section. By hovering over the blocks, we get more clear size breakdowns by method. We can double-click on these blocks to dig down deeper:

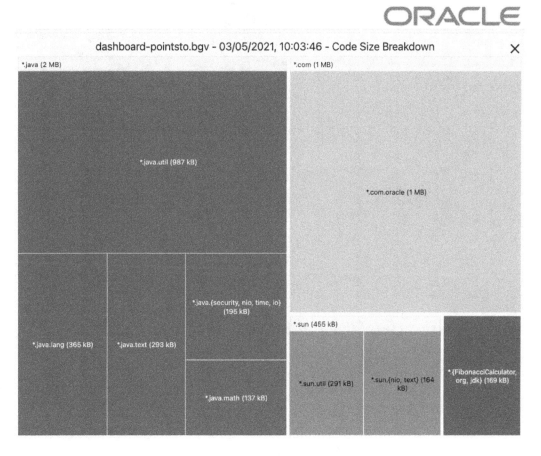

Figure 5.8 – Code size breakdown dashboard

The following screenshot shows the report we see when we double-click on the FibonacciCalculator block. Once again, we can double-click on the call graph:

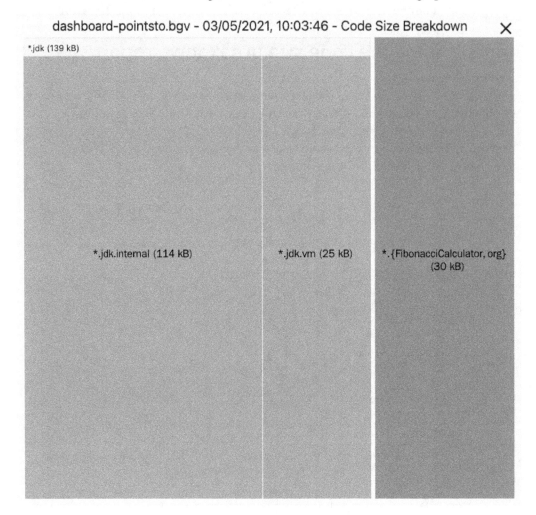

Figure 5.9 – Code size breakdown – details report

The following screenshot shows the call graph. This helps us understand the points-to analysis that the Native Image builder has performed. This can be used to identify opportunities to optimize the source code if we identify any classes' or methods' dependencies that are not used:

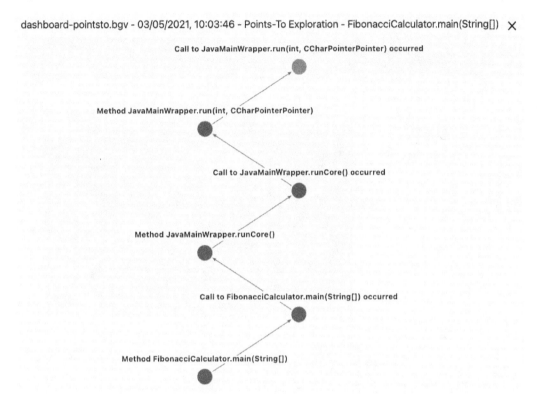

Figure 5.10 – Code points-to dependency report

Now let's look at heap size breakdown.

Heap size breakdown

Heap size breakdown provides a detailed insight into heap allocation and also provides a deep dive into the heap. We can double-click on these blocks to understand these heap allocations. The following screenshot shows the heap size breakdown report for FibonacciCalculator:

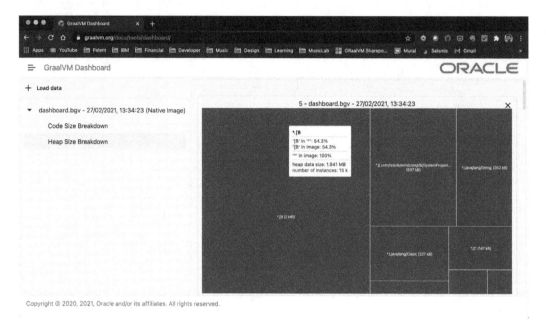

Figure 5.11 – Heap Size Breakdown dashboard

In this section, we looked at how we can use GraalVM Dashboard to analyze native images. Now let's look at how we can optimize our native images using PGO.

Understanding PGO

Using PGO, we can run the native image with an option to generate a runtime profile. The JVM creates a profile file, .iprof, which can be used to recompile the native image, to further optimize it. The following diagram (recall it from the *Profile Guided Optimization (PGO)* section in *Chapter 3, GraalVM Architecture*) shows how PGO works:

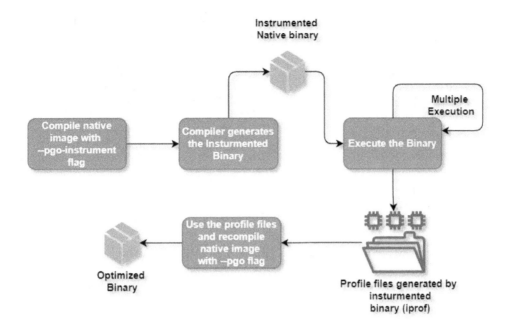

Figure 5.12 – Native Image – profile-guided optimization pipeline flow

The preceding diagram shows the native image compilation pipeline flow using PGO. Let's understand this flow better:

- The initial native image is instrumented to create a profile by passing the -pgo-instrument flag argument, while bundling the native image. This will generate a native image with instrumentation code.

- When we run the native image with several inputs, a profile is created by the native image. This profile is a file generated in the same directory with the .iprof extension.

- Once we have run through all the use cases, to ensure that the profile created covers all the paths. We can then rebuild the native image by passing the .iprof file as a parameter along with the --pgo argument.

- This will generate the optimized native image.

Let us now build an optimized native image of our FibonacciCalculator class. Let's first start by creating an instrumented native image by running the following command:

```
java -Dgraal.PGOInstrument=fibonaccicalculator.iprof -Djvmci.
CompilerIdleDelay-0 FibonacciCalculator
```

This command will build the native image using the profile information. The following screenshot shows the output of building the native image:

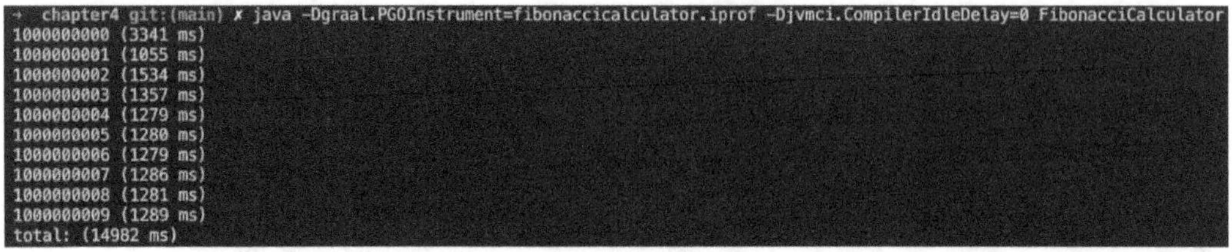

Figure 5.13 – FibonacciCalculator – generating PGO profile console output

This will generate the `fibonaccicalculator.iprof` file in the current directory. Let's now use this profile to rebuild our native image using the following command:

```
native-image -pgo=fibonaccicalculator.iprof FibonacciCalculator
```

This will rebuild the native image, with the optimum executable. The following screenshot shows the output when we build the native image with the profile:

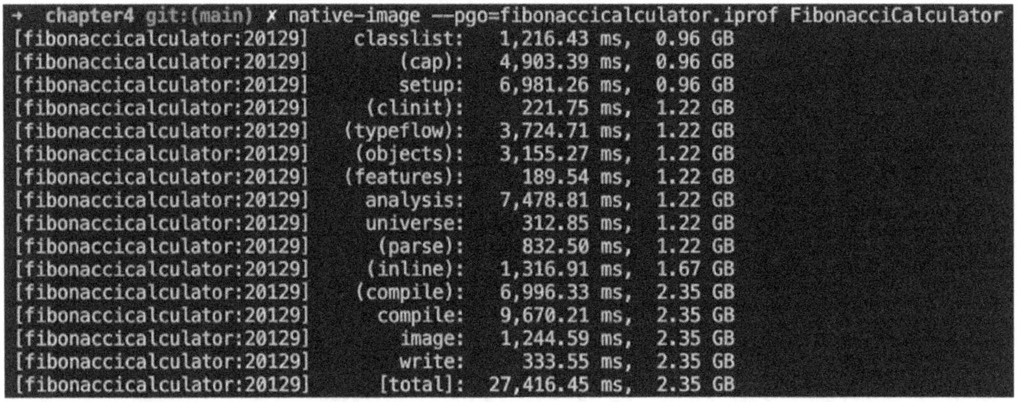

Figure 5.14 – FibonacciCalculator – generating profile-guided optimized native image console output

Now let's execute the optimized file. The following screenshot shows the output results when we run the optimized native image:

```
→  chapter4 git:(main) ✗ ./fibonaccicalculator
1000000000 (2200 ms)
1000000001 (2534 ms)
1000000002 (2905 ms)
1000000003 (2650 ms)
1000000004 (2425 ms)
1000000005 (2469 ms)
1000000006 (2486 ms)
1000000007 (2417 ms)
1000000008 (2452 ms)
1000000009 (2456 ms)
total: (24994 ms)
```

Figure 5.15 – FibonacciCalculator – running PGO image console output

As you can see, it's much faster than the original native image. Let's now compare the values. The following graph shows the compression:

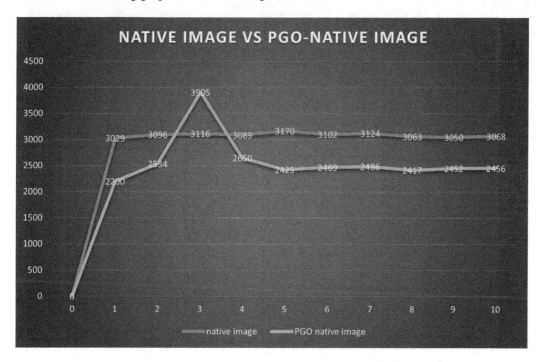

Figure 5.16 – FibonacciCalculator – native image versus PGO native image comparison

As you can see, PGO performs much faster and better.

While this is all good, if you compare this with the JIT, we see that native image does not perform so well. Let's compare this with JIT (both Graal and Java HotSpot). The following graph shows the comparison:

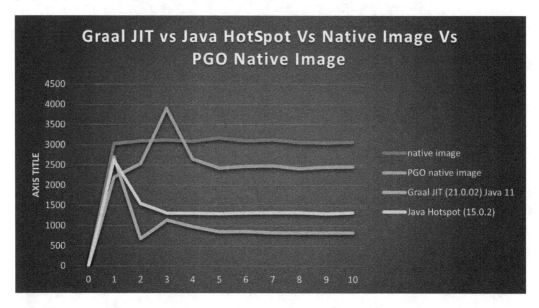

Figure 5.17 – FibonacciCalculator – Graal JIT versus Java HotSpot versus native image versus PGO native image comparison

This highlights one of the key points, that native images are not always optimum. In this case, it is definitely not, because of the heap allocation that we are doing with the large arrays that we are building. This directly impacts the performance. This is one of the areas where it's important as a developer to optimize the code. Native images use the Serial GC, hence it's not a good idea to use native images for large heaps.

Let's optimize the code and see if native images run faster than JIT. Here is the optimized code, which does the exact same logic, but uses less Heap:

```
public class FibonacciCalculator2{
        public long findFibonacci(int count) {
                int fib1 = 0;
                int fib2 = 1;
                int currentFib, index;
                long total = 0;
                for(index=2; index < count; ++index ) {
                        currentFib = fib1 + fib2;
                        fib1 = fib2;
```

```
                    fib2 = currentFib;
                    total += currentFib;
            }
            return total;
    }
    public static void main(String args[]) {
            FibonacciCalculator2 fibCal =
                new FibonacciCalculator2();
            long startTime =
                System.currentTimeMillis();
            long now = 0;
            long last = startTime;
            for (int i = 1000000000; i < 1000000010; i++)
            {
                    fibCal.findFibonacci(i);
                    now = System.currentTimeMillis();
                    System.out.printf("%d (%d
                        ms)%n", i , now - last);
                    last = now;
            }
            long endTime =
                System.currentTimeMillis();
            System.out.printf(" total: (%d ms)%n",
                    System.currentTimeMillis() -
                            startTime);
    }
}
```

Here are the final results of running this code with Graal JIT and Native Image. As you'll see in the following screenshots, Native Image does not take any time for startup and performs much faster than JIT. Let's run the optimized code with Graal JIT. Following is the output after running the optimized code:

```
→  chapter4 git:(main) ✗ java -version
java version "11.0.10" 2021-01-19 LTS
Java(TM) SE Runtime Environment GraalVM EE 21.0.0.2 (build 11.0.10+8-LTS-jvmci-21.0-b06)
Java HotSpot(TM) 64-Bit Server VM GraalVM EE 21.0.0.2 (build 11.0.10+8-LTS-jvmci-21.0-b06, mixed mode, sharing)
→  chapter4 git:(main) ✗ java FibonacciCalculator2
1000000000 (478 ms)
1000000001 (434 ms)
1000000002 (419 ms)
1000000003 (417 ms)
1000000004 (416 ms)
1000000005 (416 ms)
1000000006 (413 ms)
1000000007 (417 ms)
1000000008 (419 ms)
1000000009 (413 ms)
 total: (4242 ms)
```

Figure 5.18 – FibonacciCalculator2 – running optimized code with Graal

Now let's run the native image of the optimized code. The next screenshot shows the output after running the native image:

```
→  chapter4 git:(main) ✗ ./fibonaccicalculator2
1000000000 (0 ms)
1000000001 (1 ms)
1000000002 (0 ms)
1000000003 (0 ms)
1000000004 (0 ms)
1000000005 (0 ms)
1000000006 (0 ms)
1000000007 (0 ms)
1000000008 (0 ms)
1000000009 (0 ms)
 total: (1 ms)
```

Figure 5.19 – FibonacciCalculator2 – running optimized code as a native image

If you chart the performance, you can see a significant improvement:

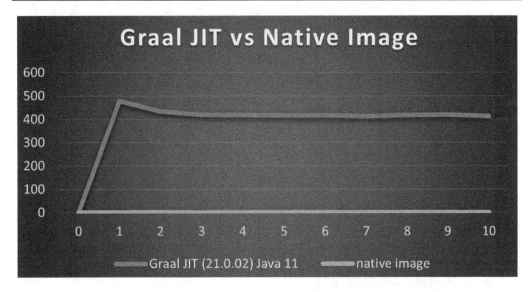

Figure 5.20 – FibonacciCalculator2 – Graal JIT versus Native Image

It's important to understand the limitations of AOT compilation and use the right approach. Let's quickly go through some of the compiler configurations for building native images.

Native image configuration

A native image build is highly configurable and it is always recommended to provide all the build configuration in the `native-image.properties` file. As the `native-image` tool takes a JAR file as an input, it is recommended to package `native-image.properties` in `META-INF/native-image/<unique-application-identifier>` within the JAR file. A unique application identifier is used to avoid any collision of resources. These paths have to be unique, as they will be configured on `CLASSPATH`. The `native-image` tool uses `CLASSPATH` to load these resources while building. Apart from `native-image.properties`, there are various other configuration files that can be packaged. We will cover some of the important configurations in this section.

The following is the typical format of the `native-image.properties` file followed by an explanation of each of the sections in the properties file:

```
Requires = <space separated list of languages that are
required>
JavaArgs = <Javaargs that we want to pass to the JVM>
Args = <native-image arguments that we want to pass>
```

- `Requires`: The `Requires` property is used to list all the language examples, such as `language:llvm language:python`.

- `JavaArgs`: We can pass regular Java arguments using this property.

- `ImageName`: This can be used to provide a custom name for the native image that is generated. By default, the native image is named after the JAR file or Mainclass file (all in lowercase letters). For example, our `FibonnaciCalculator.class` generates `fibonaccicalculator`.

- `Args`: This is the most commonly used property. It can be used to provide the native-image arguments. The arguments can also be passed from the command line, but it is much better from the configuration management perspective to have them listed in `native-image.properties`, so that it can go into Git (or any source code repository) and track any changes. The following table explains some of the important arguments that are typically used:

Command-line argument	Explanation
`--allow-incomplete-classpath`	This argument lets the native image be built with an incomplete `CLASSPATH`. The type resolution errors are reported during runtime, instead of during build time. This is normally not recommended, as we want our native image to run properly.
`--no-fallback`	This builds a standalone native image. This is the most widely used option to build a complete image, which does not depend on any other resources.
`--enable-http` and `--enable-https`	This is a very useful flag. This enables HTTP and HTTPS support respectively in the image when we are writing any HTTP server functionality.

Please refer to `https://www.graalvm.org/reference-manual/native-image/Options/` for the complete list of options.

Hosted options and resource configurations

We can configure various resources using various parameters. These resource declarations are normally configured in an external JSON file, and various `-H:` flags can be made to point to these resource files. The syntax is `-H<Resource Flag>=${.}/jsonfile.json`. The following table lists some of the important arguments that are used:

Command-line argument	Explanation
`-H:DynamicProxyConfigurationResources`	Points to the comma-separated list of JSON files that have the configuration of dynamic proxies that need to be resolved during build time and need to be configured here.
`-H:JNIConfigurationResources`	Points to the comma-separated list of JSON files that have the configuration of JNI resources that need to be resolved during build time and need to be configured here.
`-H:ReflectionConfigurationResources`	Points to the comma-separated list of JSON files that have the configuration of any classes that are used through `java.land.reflect` that need to be resolved during build time and need to be configured here.
`-H:ResourceConfigurationResources`	Points to the JSON file that has the configuration of any external resources that are used by the application that need to be resolved during build time.
`-H:SubstitutionResources`	This is a comma-separated list of resource filenames with declarative substitutions.

Command-line argument	Explanation
`-H:SerializationConfigurationResources`	This is a comma-separated list of resource files that have configuration classes for serialization. This is new support that was added in 21.0 and above. Using this support, we can serialize objects. To configure which classes support this, the Native Image builder needs to know at build time. These classes need to implement `java.io.Serializable`.
`-H:Class`	The class containing the default entry point method. Ignored if `kind != EXECUTABLE`.
`-H:Name`	Name of the output file to be generated.
`-H:IncludeResources`	Resources to be included. This is a pipe (`|`) separated list of resources.
`-H:+PrintClassInitialization`	Prints the class initialization comments while building. This helps us track the class initializations and helps us identify the classes that we want to initialize at build time or runtime using the `--initialize-at-build-time` and `--initialize-at-run-time` flags
`-H:Dump:`	This flag is used to create Graal graphs of the native image.

`native-images.properties` captures all the configuration parameters, and it is a good practice to pass configurations through the `native-image.properties` file, as it's easy to manage it in a source code configuration management tool.

GraalVM comes with an agent that tracks the dynamic features of a Java program at runtime. This helps in identifying and configuring the native image build with dynamic features. To run the Java application with the Native image agent, we need to pass `-agentlib:native-image-agent=config-output-dir=<path to config dir>`. The agent tracks the execution and intercepts calls that look up classes, methods, resources, and proxies. The agent then generates `jni-config.json`, `reflect-config.json`, `proxy-config.json`, and `resource-config.json` in the config directory passed as the parameter. It's a good practice to run the application multiple times, with different test cases, to ensure that the complete code is covered, and the agent gets to catch most of the dynamic calls. When we run iterations, it is important to use `-agentlib:native-image-agent=config-merge-dir=<path to config dir>` so that the configuration files are not overwritten but merged.

We can generate Graal graphs with native images, to analyze how the native image is running. In the next section, we will explore how to generate these Graal graphs.

Generating Graal graphs for native images

Graal graphs can be generated even for native images. Graal graphs can be generated during build time or at runtime. Let's explore this feature in this section using our `FibonnaciCalculator` application.

Let's generate the dump for `FibonacciCalculator` using this command:

```
native-image -H:Dump=1 FibonacciCalculator
```

The following is the output:

```
native-image -H:Dump=1 FibonacciCalculator

[fibonaccicalculator:54143]    classlist:   811.75 ms,   0.96 GB
[fibonaccicalculator:54143]        (cap):  4,939.64 ms,   0.96 GB
[fibonaccicalculator:54143]        setup:  6,923.28 ms,   0.96 GB
[fibonaccicalculator:54143]     (clinit):   155.70 ms,   2.29 GB
[fibonaccicalculator:54143]    typeflow):  3,841.07 ms,   2.29 GB
[fibonaccicalculator:54143]     (objects):  3,235.92 ms,   2.29 GB
[fibonaccicalculator:54143]   (features):   169.55 ms,   2.29 GB
[fibonaccicalculator:54143]     analysis:  7,550.64 ms,   2.29 GB
[fibonaccicalculator:54143]     universe:   295.75 ms,   2.29 GB
[fibonaccicalculator:54143]       (parse):   829.58 ms,   3.18 GB
[fibonaccicalculator:54143]      (inline):  1,357.72 ms,   3.18 GB
```

```
[Use -Dgraal.LogFile=<path> to redirect Graal log output
to a file.]
Dumping IGV graphs in /graal_dumps/2021.02.28.20.32.24.880
Dumping IGV graphs in /graal_dumps/2021.02.28.20.32.24.880
[fibonaccicalculator:54143]    (compile):  13,244.17 ms, 4.74 GB
[fibonaccicalculator:54143]      compile:  16,249.17 ms, 4.74 GB
[fibonaccicalculator:54143]        image:   1,816.98 ms,  4.74 GB
[fibonaccicalculator:54143]        write:     430.54 ms,     4.74 GB
[fibonaccicalculator:54143]      [total]:  34,247.73 ms, 4.74 GB
```

This command generates a lot of graphs for every class that is initialized. We can use the -H:MethodFilter flag to specify the class and method that we want to generate a graph for. The command would look something like this:

```
native-image -H:Dump=1 -H:MethodFilter=FibonacciCalculator.main
FibonacciCalculator
```

Please refer to the section *Graal intermediate representation* from *Chapter 4, Graal Just-In-Time Compiler*, to know how to read these graphs and understand opportunities for optimizing the code. Optimizing the source code is critical for native images as there are no runtime optimizations like we have in just-in-time compilers.

Understanding how native images manage memory

Native images come bundled with Substrate VM, which has the functionality of managing memory, including garbage collection. As we saw in the *Building native images* section, the heap allocation happens as part of the image creation to speed up the startup. These are classes that are initialized at build time. Refer to *Figure 5.3* to see how the Native Image builder initialized the heap region after performing static region analysis. At runtime, a garbage collector manages the memory. There are two garbage collection configurations that the Native Image builder supports. Let's understand these two garbage collection configurations in the following subsections.

The Serial garbage collector

The Serial Garbage Collector (GC) is the default that gets built into the native image. This is available both on the Community and Enterprise edition. This garbage collector is optimized for a low memory footprint and small heap size. We can use the --gc=serial flag to explicitly use the Serial GC. The Serial GC is a simple implementation of the GC.

The Serial GC divides the heap into two regions, that is, young and old. The following figure shows how the Serial GC works:

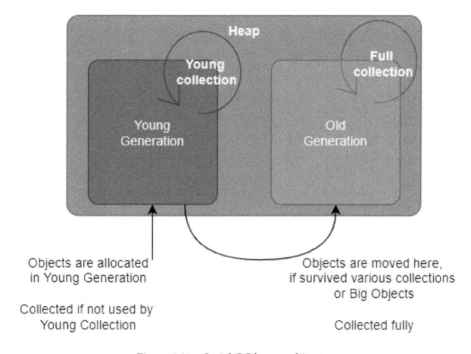

Figure 5.21 – Serial GC heap architecture

The young generation is used for new objects. It is triggered when the young generation block is full and all the objects that are not used are reclaimed. When the old generation blocks become full, a full collection is triggered. Young collections run faster and a full collection is more time-consuming at runtime. This behavior can be tweaked using the argument -XX:PercentTimeInIncrementalCollection.

By default, this percentage is 50. This can be increased to reduce the number of full collections, improving the performance, but will have a negative effect on the memory size. Depending on the memory profiling, while testing the application, we can optimize this parameter for better performance and memory footprint. Here is an example of how to pass this parameter at runtime:

```
./fibonaccicalculator -XX:PercentTimeInIncrementalCollection=40
```

This parameter can also be passed at build time:

```
native-image --gc=serial
 -R:PercentTimeInIncrementalCollection=70 FibonacciCalculator
```

There are other arguments that can also be used for fine-tuning, such as -XX:MaximumYoungGenerationSizePercent. This argument can be used to tweak the maximum percentage the young generation block should occupy of the overall heap.

The Serial GC is single-threaded and works well for small heaps. The following figure shows how the Serial GC works. The application threads are paused to reclaim the memory. It is called a *Stop the World* event. During this time, the **Garbage Collector Thread** runs and reclaims the memory. This has an impact on the performance of the application if the heap size is large and if there are a lot of threads running. The Serial GC is very good for small processes with small heap sizes.

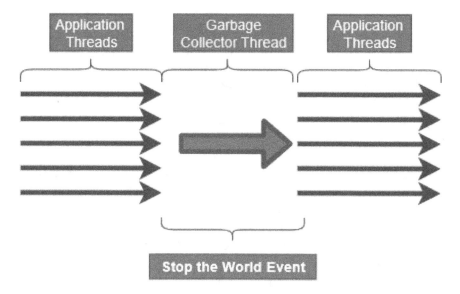

Figure 5.22 – Serial GC heap flow

By default, the Serial GC assumes an 80% heap size, before it starts the GC thread. This can be changed using the -XX:MaximumHeapSizePercent flag. There are other flags that can be used to fine-tune the performance of the Serial GC.

The G1 Garbage Collector

The G1 Garbage Collector is the more recent and advanced implementation of the garbage collector. This is only available in the Enterprise edition. The G1 Garbage Collector can be enabled using the flag --gc=G1. G1 provides the right balance of throughput and latency. Throughput is the average time spent running the code versus GC. Higher throughput means we have more CPU cycles for code, rather than the GC thread. Latency is the amount of time the *Stop The World* event takes or the time taken to pause the code execution. The less latency, the better for us. G1 targets high throughput and low latency. Here is how it works.

G1 divides the whole heap into small regions. G1 runs concurrent threads to find all live objects, and the Java application is never paused, and keeps track of all the pointers between regions, and tries to collect regions so that there are shorter pauses in the program. G1 might also move live objects and consolidate them into regions and tries to make the regions empty.

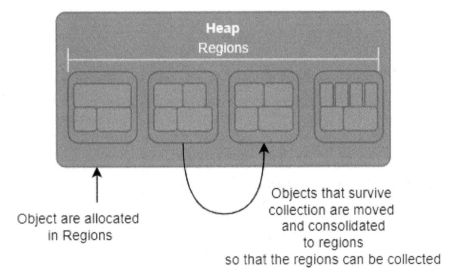

Figure 5.23 – G1 GC heap flow

The previous diagram shows how G1 GC works by dividing into regions. The allocation of objects into the regions is based on trying to allocate memory in empty regions and trying to empty the regions by consolidating the objects into regions, like partitioning and de-partitioning. The idea is to optimally manage and collect the regions.

G1 garbage collector has a larger footprint than the Serial GC and is for longer-running, larger heap sizes. There are various parameters that can be used to fine-tune the performance of G1 GC. Listed next are some of them (`-H` are the parameters passed during the building of the image and `-XX` is passed while running the image):

- `-H:G1HeapRegionSize`: This is the size of each region.

- `-XX:MaxRAMPercentage`: Percentage of physical memory size that is used as the heap size.

- `-XX:ConcGCThreads`: Number of concurrent GC threads. This needs to be optimized for the best performance.

- `-XX:G1HeapWastePercent`: The garbage collector stops claiming when it reaches this percentage. This will allow lower latency and higher throughput, however, it is critical to set an optimum value, as if it's too high, then the objects will never get collected, and the application memory footprint will always be high.

The choice of using the right garbage collector and the configuration is critical for the performance and memory footprint of the application.

Managing the heap size and generating heap dumps

Heap size can be manually set using the following runtime parameters passed to the native image while running them. `-Xmx` sets the maximum heap size, `-Xms` sets the minimum heap size, and `-Xmn` sets the size of the young generation region, in bytes. Here is an example of how these arguments can be used at runtime:

```
./fibonaccicalculator -Xms2m -Xmx10m -Xmn1m
```

At build time, we can pass arguments to configure the heap size. This is a critical configuration and has to be done with a lot of care as this has a direct impact on the memory footprint and performance of the native image. The following command is an example that configures the minimum heap size, maximum heap size, and the maximum new size of the heap:

```
native-image -R:MinHeapSize=2m -R:MaxHeapSize=10m
 -R:MaxNewSize=1m FibonacciCalculator
```

Heap dumps are the most important for debugging any memory leaks and memory management issues. We typically use tools such as VisualVM to do such heap dump analysis. Native images are not built with **Java Virtual Machine Tool Interface (JVMTI)** agents to perform heap dumps on while the application is running. However, while building the native image, we can build with the `-H:+AllowVMInspection` flag. This will create a native image that can generate a stack dump when we send a USR1 signal (`sudo kill -USR1` or `-SIGUSR1` or QUIT/BREAK keys) and a runtime compilation information dump when we send a USR2 signal (`sudo kill -USR2` or `-SIGUSR2` – you can check the exact signal using the `kill -l` command). This feature is available only in the Enterprise edition.

We can also programmatically create heap dumps by calling `org.graalvm.nativeimage.VMRuntime#dumpHeap` when required.

Building static native images and native shared libraries

Static native images are statically linked binaries that do not need any additional dependent libraries at runtime. These are very useful when we are building microservice applications as native images so that they can be easily packaged into Docker, without worrying about dependencies. Static images are best for building container-based microservices.

At the time of writing this book, this feature is only available for Linux AMD64 on Java 11. Please refer to `https://www.graalvm.org/reference-manual/native-image/StaticImages/` for the latest updates and the process of building static native images.

The Native Image builder also builds shared libraries. Sometimes you may want to create your code as a shared library that is used by some other application. For that, you have to pass the `-shared` flag to build a shared library, instead of an executable one.

Debugging native images

Debugging a native image requires building the image with debug info. We can use `-H:GenerateDebugInfo=1`. Here is an example of using this argument for FibonnacciCalculator:

```
native-image -H:GenerateDebugInfo=1 FibonacciCalculator
```

The generated image has debug information in the form of the **GNU Debugger** (**GDB**). This can be used to debug the code at runtime. The following shows the output of running the preceding command:

```
native-image -H:GenerateDebugInfo=1 FibonacciCalculator
```

```
[fibonaccicalculator:57833]    classlist:   817.01 ms,    0.96 GB
[fibonaccicalculator:57833]        (cap):  6,301.03 ms,    0.96 GB
[fibonaccicalculator:57833]        setup:  9,946.35 ms,    0.96 GB
[fibonaccicalculator:57833]     (clinit):   147.54 ms,    1.22 GB
[fibonaccicalculator:57833]   (typeflow):  3,642.34 ms,    1.22 GB
[fibonaccicalculator:57833]    (objects):  3,164.39 ms,    1.22 GB
[fibonaccicalculator:57833]   (features):   181.00 ms,    1.22 GB
[fibonaccicalculator:57833]     analysis:  7,282.44 ms,    1.22 GB
[fibonaccicalculator:57833]     universe:   304.43 ms,    1.22 GB
[fibonaccicalculator:57833]      (parse):   624.60 ms,    1.22 GB
[fibonaccicalculator:57833]     (inline):   989.65 ms,    1.67 GB
[fibonaccicalculator:57833]    (compile):  8,486.97 ms,    3.15 GB
[fibonaccicalculator:57833]      compile: 10,625.01 ms,    3.15 GB
[fibonaccicalculator:57833]        image:   869.81 ms,    3.15 GB
[fibonaccicalculator:57833]    debuginfo:  1,078.95 ms,    3.15 GB
[fibonaccicalculator:57833]        write:  2,224.22 ms,    3.15 GB
[fibonaccicalculator:57833]      [total]: 33,325.95 ms,    3.15 GB
```

This will generate a `sources` directory, which holds the cache that is generated by the native image builder. This cache brings JDSK, GraalVM, and application classes to help with debugging. The following is the output of listing the contents of the `sources` directory:

```
$ ls -la
total 8
drwxr-xr-x     9 vijaykumarab     staff      288 17 Apr 09:01 .
drwxr-xr-x    10 vijaykumarab     staff      320 17 Apr 09:01 ..
-rw-r--r--     1 vijaykumarab     staff     1240 17 Apr 09:01
FibonacciCalculator.java
drwxr-xr-x     3 vijaykumarab     staff       96 17 Apr 09:01
com
drwxr-xr-x     5 vijaykumarab     staff      160 17 Apr 09:01
java.base
```

```
drwxr-xr-x      3 vijaykumarab     staff        96 17 Apr 09:01
jdk.internal.vm.compiler
drwxr-xr-x      3 vijaykumarab     staff        96 17 Apr 09:01
jdk.localedata
drwxr-xr-x      3 vijaykumarab     staff        96 17 Apr 09:01
jdk.unsupported
drwxr-xr-x      3 vijaykumarab     staff        96 17 Apr 09:01
org.graalvm.sdk
```

To debug native images, we need the gdb utility. Please refer to https://www.gnu.org/software/gdb/ for how to install gdb for your target machine. Once properly installed, we should be able to enter the gdb shell by executing the gdb command. The following shows the typical output:

```
$ gdb
GNU gdb (GDB) 10.1
Copyright (C) 2020 Free Software Foundation, Inc.
License GPLv3+: GNU GPL version 3 or later <http://gnu.org/
licenses/gpl.html>
This is free software: you are free to change and redistribute
it.
There is NO WARRANTY, to the extent permitted by law.
Type "show copying" and "show warranty" for details.
This GDB was configured as "x86_64-apple-darwin20.2.0".
Type "show configuration" for configuration details.
For bug reporting instructions, please see:
<https://www.gnu.org/software/gdb/bugs/>.
Find the GDB manual and other documentation resources online
at:
        <http://www.gnu.org/software/gdb/documentation/>.

For help, type "help".
Type "apropos word" to search for commands related to "word".
(gdb)
```

We need to point to the directories where we generated the source files in the previous step. We can do that by executing the following command:

```
set directories /<sources directory>/jdk:/ <sources directory>/
graal:/ <sources directory>/src
```

Once the environment is set, we can use `gdb` to set the breakpoints and debug. Please refer to `https://www.gnu.org/software/gdb/documentation/` for detailed documentation on how to use `gdb` to debug executables.

At the time of writing this book, debug information can be used to perform breakpoints, single-stepping, stack backtrace, printing primitive values, the casting and printing of objects, path expressions, and references by method name and static data. Please refer to `https://www.graalvm.org/reference-manual/native-image/DebugInfo/` for more details.

Limitations of Graal AOT (Native Image)

In this section, we will go through some of the limitations of Graal AOT and native images.

Graal ahead-of-time compilation performs static analysis with the closed-world assumption. It assumes that all the classes that are reachable at runtime are available during build time. This has a direct implication on writing any code that requires dynamic loading – such as Reflection, JNI, Proxies, and so on. However, the Graal AOT compiler (`native-image`) provides a way to provide this metadata in the form of JSON manifest files. These files can be packaged along with the JAR file, as an input for the compiler:

- Loading classes dynamically: Classes that are loaded at runtime, which will not be visible to the AOT compiler at build time, need to be specific in the configuration file. These configuration files are typically saved under `META-INF/native-image/`, and should be in `CLASSPATH`. If the class is not found during the compilation of the configuration file, it will throw a `ClassNotFoundException`.

- Reflection: Any call to the `java.lang.reflect` API to list the methods and fields or invoke them using the reflection API has to be configured in the `reflect-config.json` file under `META-INF/native-image/`. The compiler tries to identify these reflective elements through static analysis.

- Dynamic Proxy: Dynamic proxy classes that are generated instances of `java.lang.reflect.Proxy` need to be defined during build time. The interfaces need to be configured in proxy-config.json.

- **Java Native Interface** (JNI): Like reflection, JNI also accesses a lot of class information dynamically. Even these calls need to be configured in `jni-config.json`.

- Serialization: Java serialization also accesses a lot of class metadata dynamically. Even these accesses need to be configured ahead of time.

You can find more details about the other limitations here: `https://www.graalvm.org/reference-manual/native-image/Limitations/`.

GraalVM containers

GraalVM also comes packaged as a Docker container. It can be directly pulled from the Docker Registry (`ghcr.io`) or can be used as a base image to build custom images. Here are some of the key commands to use GraalVM containers:

- To pull the Docker image: `docker pull ghcr.io/graalvm/graalvm-ce:latest`

- To run the container: `docker run -it ghcr.io/graalvm/graalvm-ce:latest bash`

- To use in the Dockerfile as a base image: `FROM ghcr.io/graalvm/graalvm-ce:latest`

We will be exploring more about GraalVM containers in *Chapter 9, GraalVM Polyglot – LLVM, Ruby, and WASM*, when we talk about building microservices on GraalVM.

Summary

In this chapter, we went through Graal just-in-time and ahead-of-time compilers in detail. We took sample code and looked at how Graal JIT performs various optimizations. We also went through, in detail, how to understand Graal graphs. This is critical knowledge that will help in analyzing and identifying optimizations that we can do during development, to speed up Graal JIT compilation at runtime.

The chapter provided detailed instructions on how to build native images, and how to optimize native images using profile-guided optimization. We took sample code and compiled native image, and also found out how a native image works internally. We identified code issues that might cause native images to run slower than just-in-time compilers. We also covered the limitations of native images, and when to use native images. We explored various build time and runtime configurations to optimize a build and running native images.

In the next chapter, we will get into understanding the Truffle language implementation framework and how to build polyglot applications.

Questions

1. How are native images created?

2. What is points-to analysis?

3. What is region analysis?

4. What are the Serial GC and the G1 GC?

5. How do you optimize native images? What is PGO?

6. What are the limitations of native images?

Further reading

- GraalVM Enterprise edition (`https://docs.oracle.com/en/graalvm/enterprise/19/index.html`)

- Graal VM Native Image documents (`https://www.graalvm.org/reference-manual/native-image/`)

Section 3: Polyglot with Graal

This section explains how we can use GraalVM as a polyglot VM, with a hands-on session on how polyglot interoperability works and how Truffle helps with the polyglot support of Java, Python, and JavaScript. This section includes the following chapters:

- *Chapter 6, Truffle for Multi-language (Polyglot) support*
- *Chapter 7, GraalVM Polyglot – JavaScript and Node.js*
- *Chapter 8, GraalVM Polyglot – Java on Truffle, Python, and R*
- *Chapter 9, Graal Polyglot – LLVM, Ruby, and WASM*

6
Truffle for Multi-language (Polyglot) support

Support for polyglot development is one of the biggest features of GraalVM. In *Chapter 4*, *Graal Just-In-Time Compiler*, and *Chapter 5*, *Graal Ahead-of-Time Compiler and Native Image*, we went into a lot of detail on how Graal optimizes code, both at build time and run time. We have only used Java in all the previous chapters. However, GraalVM extends most of its advanced features to other programming languages too. GraalVM provides a language implementation framework called **Truffle Language Implementation Framework** (commonly known as **Truffle**).

GraalVM not only provides a high-performance runtime for JVM languages such as Java, Groovy, Kotlin, and Scala, but it also supports non-JVM languages such as JavaScript, Ruby, Python, R, WebAssembly, and LLVM languages that implement Truffle. A lot more languages are being implemented on Truffle.

This chapter provides a conceptual view of how Truffle helps guest language developers and provides a well-designed, high-performance framework to build applications using guest languages on top of GraalVM. This chapter does not get into too much detail on how to write in guest languages using Truffle. This is only intended to describe Truffle's architecture and concepts at a high level so that you can follow the subsequent chapters on how non-JVM languages are implemented on GraalVM.

In this chapter, we will cover the following topics:

- Exploring the Truffle language implementation framework
- Exploring the Truffle interpreter/compiler pipeline
- Learning Truffle DSL
- Understanding how Truffle supports interoperability
- Understanding Truffle instrumentation
- Ahead-of-time compilation using Truffle
- Optimizing Truffle interpreter performance with launcher options
- SimpleLanguage and Simple Tool

By the end of this chapter, you will have a good understanding of Truffle's architecture, and how Truffle provides a framework for other programming languages to run on GraalVM.

Exploring the Truffle language implementation framework

In *Chapter 3*, *GraalVM Architecture*, in the Truffle section, we briefly covered the architecture of Truffle. Truffle is an open source library that provides a framework to implement language interpreters. Truffle helps run guest programming languages that implement the framework to utilize the Graal compiler features to generate high-performance code. Truffle also provides a reference implementation called SimpleLanguage to guide developers to write interpreters for their languages. Truffle also provides a tools framework that helps integrate and utilize some of the modern diagnostic, debugging, and analysis tools.

Let's understand how Truffle fits into the overall GraalVM ecosystem. Along with interoperability between the languages, Truffle also provides embeddability. Interoperability allows the calling of code between different languages, while embeddability allows the embedding of code written in different languages in the same program.

Language interoperability is critical for the following reasons:

- Different programming languages are built to solve different problems, and they come with their own strengths. For example, we use Python and R extensively for machine learning and data analytics, and we use C/C+ for high-performance mathematical operations. Imagine if we would reuse the code as it is, either by calling the code from a host language (such as Java) or embedding that code within the host language. This also increases the reusability of the code and allows us to use an appropriate language for the task at hand, rather than rewriting the logic in different languages.

- Large migration projects where we are moving from one language to another can be phased out if we have the feature of multiple programming language interoperability. This brings down the risk of migration considerably.

The following figure illustrates how to run applications written in other languages on GraalVM:

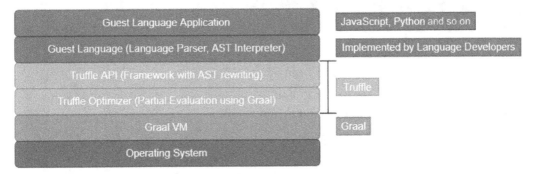

Figure 6.1 – Truffle stack

In the figure, we can see GraalVM, which is the JVM and Graal JIT compiler that we covered in the previous chapters. On top of that, we have the Truffle framework. Truffle has two major components. They are as follows:

- **Truffle API**: The Truffle API is the language implementation framework that any guest language programmers can use to implement the Truffle interpreter for their respective languages. Truffle provides a sophisticated API for **Abstract Syntax Tree (AST)** rewriting. The guest language is converted to AST for optimizing and running on GraalVM. The Truffle API also helps in providing an interoperability framework between languages that implement the Truffle API.

- **Truffle optimizer**: The Truffle optimizer provides an additional layer of optimization for speculative optimization with partial evaluation. We will be going through this in more detail in the subsequent sections.

Above the Truffle layer, we have the guest language. This is JavaScript, R, Ruby, and others that implement the Truffle Language Implementation framework. Finally, we have the application that runs on top of the guest language runtime. In most cases, application developers don't have to worry about changing the code to run on GraalVM. Truffle makes it seamless by providing a layer in between. The following diagram shows a detailed stack view of the GraalVM and Truffle ecosystem:

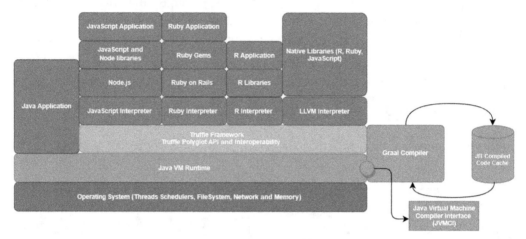

Figure 6.2 – Truffle and Graal detailed stack view

This diagram is a simple representation of how Truffle acts as a layer between non-JVM languages and GraalVM. Let's understand this in detail.

Truffle provides the API that the individual interpreters implement to rewrite the code into ASTs. The AST representation is later converted to a Graal intermediate representation for Graal to execute and also optimize just in time. The guest languages run on top of the Truffle interpreter implementations of the respective guest languages.

Let's look at how these various layers interact and how Truffle helps the guest language to run on GraalVM.

Exploring the Truffle interpreter/compiler pipeline

Truffle provides a Java library that can be used to write AST interpreters in Java for any language. The guest language semantics are expressed as an AST using the AST interpreters. Graal and Truffle specify the exact format of AST, and the framework enforces this specification. So, any guest language AST interpreter written using Truffle Language Implementation API will generate the semantics of the language in AST that GraalVM can then use to optimize just in time and run. The following figure provides a detailed flow of how Truffle and Graal work:

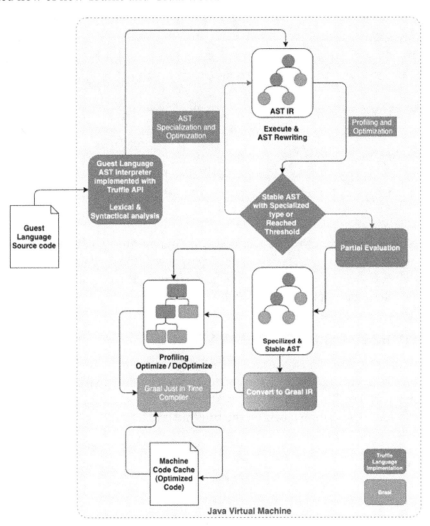

Figure 6.3 – Truffle and Graal compilation pipeline

Let's understand the flow diagram better.

The guest languages are parsed using a traditional syntactical and lexical analyzer. This generates an AST. The AST is an intermediate representation built as per the Truffle/Graal specifications. The AST is a tree structure where each parent node is an operation/function, and the child nodes are data/operands/dependencies. Each node that is part of the AST has an `execute()` method. The `execute()` method implements the semantics of the corresponding language construct. The guest language developers are supposed to design these nodes and provide facilities to parse the guest language code and build the tree. The following code snippet shows the abstract node object:

```
abstract class Node {
    // Executes the operation encoded by this
    // node and returns the result.
    public abstract Object execute(Frame f);


    // Link to the parent node and utility to
    // replace a node in the AST.
    private Node parent;


    protected void replace(Node newNode);
}
```

All the node types implement this abstract node. AST interpreters go through the AST node by node to execute using a stack frame to keep track of the execution and data. AST interpreters are easy to implement but have huge performance overheads. Let's now look at how Truffle interpreters optimize the AST by rewriting the tree.

Self-optimization and tree rewriting

The structure of the AST can be dynamically changed by rewriting. This is also sometimes referred to as AST rewriting. The AST interpreter can rewrite the AST dynamically based on the runtime profiling. This improves the performance significantly. The nodes can be updated at runtime by rewriting based on the runtime profiling and optimizing the AST.

One of the biggest issues with dynamic languages, from a performance perspective, is the type declarations (unlike languages such as C, where the type is specified in the code). The types are not declared at coding time, which means the variables, at the start of the execution, may assume any type. The profiler can figure out the specific type after a few runs and can then specialize the type of the variables based on the runtime profiling. It then rewrites the AST with the specific type. This technique is called **Type Specialization**. This has a significant performance boost, as the operations/methods that are performed on this data can be specialized to the data type. The Truffle interpreter leaves a speculation *guard* in case the speculation/assumption made on the type is proven wrong in future executions. In this case, the guard is invoked to deoptimize the AST.

Tree rewriting also happens when an operation/function has to be resolved at runtime. This is very important for the dynamic resolution of function or operations, due to polymorphic implementations of the operations or functions. When the profiler identifies a specific implementation of the function/operation, the resolved function/operation can be rewritten. A simple diagram that illustrates operation resolution follows this paragraph. The `Node` class has the `replace()` method, which is used for rewriting. The following figure illustrates a very simple expression, and shows how the AST gets rewritten after profiling:

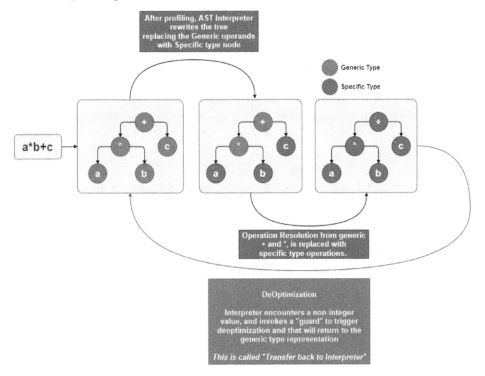

Figure 6.4 – AST specialization illustration

On the left side of the figure, the Truffle interpreter starts off as a generic AST node. Based on the profiler, the interpreter understands that it's mostly a specific type value, let's say an integer. It will rewrite the nodes with integer types and the operation nodes with integer operations. This optimizes the execution of the AST. In the left-most representation, + might mean the concatenation of strings or the addition of integers or longs or floats, or any other polymorphic execution of the operation that is reduced to the right-most representation, where it is very clearly an integer. A guard check is placed there in case the assumption made is proven wrong in the future. If the operands happen to be floats, or some other type, then a deoptimization is invoked, which might take it back to the left-most representation. The Truffle interpreter will once again profile to identify the right type specializations to apply. It may identify that it is more of a long or double, so it might rewrite the AST as a double and optimize. Type specialization is applied to local variables, return types, operations, and fields.

AST rewriting based on type specialization provides a significant boost to performance, and by the time we get to more advanced optimizations that the Graal performs, we have a very stable AST.

Since the AST interpreters are implemented in Java, the node `execute()` method is written to handle a generic object. Type specialization also helps by replacing the `execute()` method with a specialized `executeInt()` method, and also reduces the load on the CPU for boxing and unboxing by replacing the wrapper implementations (`Integer`, `Double`, and so on) with primitive types (`int`, `double`). This technique is sometimes referred to as *boxing elimination*.

Once the interpreter finds that there are no node rewrites, that means that the AST has been stable. The code is then compiled to machine code, and all the virtual calls are inlined with specific calls. This code is then passed to the Graal JIT compiler for further optimization at runtime. This is called **partial evaluation**. Guards are embedded into the code in case the assumptions made are valid. When any of the assumptions made are invalid, a guard code will bring back the execution to the AST interpreter, where node rewriting will once again happen. This is called **Transfer to Interpreter**.

Partial Evaluation

The calls to these `execute()` methods are virtual dispatches, which have a significant overhead on the performance. As Truffle identifies the code that is stabilized, when no more AST rewriting happens and the code has a lot of calls, it performs partial evaluation to improve the performance of the execution of this code. Partial evaluation includes inlining the code, eliminating the virtual dispatches and replacing them with direct calls, and building a combined unit that will be submitted to the Graal compiler for further optimization. Truffle places guard points wherever the assumptions made might be disproved. These guard points trigger the deoptimization by invalidating the code and switching back to interpreter mode of execution. The code is then compiled to machine code for the guest language after aggressive constant folding, inlining, and escape analysis. Truffle performs inlining on the AST so that it is language agnostic. The inlining decisions are taken by performing partial evaluation on every candidate.

In the next section, we'll have a look at the Truffle framework in GraalVM, which is used to create DSLs.

Learning Truffle DSL

Truffle defines a **Domain-Specific Language** (**DSL**) based on the Java annotation processor. The language developer has to write a lot of boilerplate code to manage the states of the specializations. To appreciate how Truffle DSL makes a programmer's life easy, let's take a quick example:

```
c = a + b
```

As we discussed earlier in this chapter, in AST, every operation and operand is represented as a node. In Truffle, it is a Java class derived from `com.oracle.truffle.api.nodes.Node`. To understand the need for a DSL, let's oversimplify the implementation of AST for the preceding expression.

Since we are looking at dynamically typed languages, a and b can be any type. We need an expression node that should implement an `execute` method, which checks for all the possible types for a and b. We will have to write logic something like this:

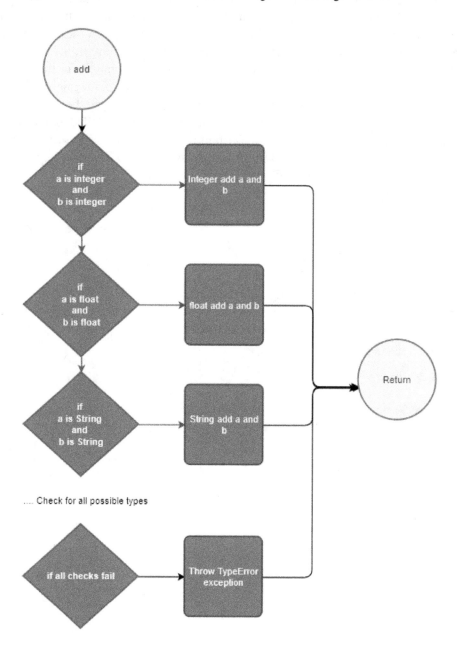

Figure 6.5 – Guard checks for implementing specialization – flow chart

In the preceding flow chart, we are checking for all possible combinations of operand types and evaluating the expression based on that, and if none of these conditions satisfy, it throws a `TypeError` exception. This kind of logic needs to be written in the Truffle interpreter, as we are dealing with dynamically typed languages.

If we convert this to Truffle interpreter code, this is a lot of code for a very simple expression. Imagine if we have more complex expressions and other operations and functions. The Truffle interpreter code will be a nightmare to write and manage.

This is where Truffle DSL solves the issue. Truffle DSL provides a very well-defined framework of node hierarchy, annotations, and annotation processors that can be used to handle this type of dynamism.

The `@Specialization` annotation is a specialization annotation implemented by the `com.oracle.truffle.api.dsl.Specialization` class, and this is used as an annotation for all the possible evaluation cases (green boxes in the previous figure). Truffle DSL compiles this into a dynamic code where Truffle picks the right implementation (the first in the sequence) based on the operand parameters. The language developer code will look something like the following code snippet:

```
@Specialization protected long executeAddInt (int left, int
right) {
    return left + right;
}
@Specialization String executeAddFloat (Float left, Float
right) {
    return left + right;
}
@Specialization String executeAddString (String left, String
right) {
    return left + right;
}
```

The preceding code shows how Truffle DSL simplifies the job, and there is no need to write a lot of `if`/`else` statements. Truffle DSL annotations take care of compiling and generating that code for us. And finally, to handle the exception case, we can use the `@Fallback` annotation implemented by the `com.oracle.truffle.api.dsl.Fallback` class. The fallback code block will look something like the following code snippet:

```
@Fallback protected void typeError (Object left, Object right)
{
```

```
    throw new TypeException("type error: args must be two
    integers or floats or two", this);
}
```

As mentioned before, Truffle picks the right implementation based on the operand types dynamically, by default. However, this can be also modified by declaring guards with the @Specialization annotation. Four types of guard can be declared for the @Specialization annotation. They are as follows:

- **Type guard**: Type guards pick the specialized method based on the parameter types. The parameter values are checked for the type and if the operand (or child nodes annotated as @NodeChild in the Node class declaration) types match, then that particular method is executed.

- **Expression guard**: The language developer can declare specific custom expressions in the @Specification annotation. The expressions are very simple Java-like code that evaluates to a Boolean value. If this expression is evaluated to true, then that particular method is executed; if it is false, the interpreter skips that execution. Here is a simple example:

```
@Specialization(guards = {"!isInteger(operand)",
    "!isFloat(operand)"})
protected final int executeTheMethod(final Object
operand) {
    //....code to execute if the expression is true
}
```

In the preceding code, the executeTheMethod() method gets picked by the Truffle interpreter if the expression that is passed in guards is true. In this case, it will be true if the operand is not an integer and is not a float. guards is actually a String array attribute in com.oracle.truffle.api.dsl.Specialization. We can pass multiple expressions.

- **Event guard**: An event guard is a powerful way to handle exception cases. Let's assume in the flowchart depicted in *Figure 6.5* we have an evaluation that might throw an exception, such as ArthimeticException. We could have multiple specialization implementations to rewrite that execution to handle exception cases. To understand this better, let's look at the following code example:

```
@Specialization(rewriteOn = ArithmeticException.class)
int executeNoOverflow(int a, int b) {
    return Math.addExact(a, b);
}
```

```
@Specialization
long executeWithOverflow(int a, int b) {
    return a + b;
}
```

In this code, Truffle will call the `executeWithOverflow()` method when integer types are matching (type guard), but if the integer values cause an overflow, `ArthimeticException` is thrown. In that case, Truffle will use the `executeNoOverflow()` method to overwrite the add method. This is an example of node rewriting based on specialization, which we discussed previously in this chapter.

- **Assumption guard**: An `Assumption` object is used by Truffle to validate and invalidate an assumption where `com.oracle.truffle.api.Assumption` is an interface. Once an assumption is invalidated, it can never be valid in that runtime. This is used by Truffle to take decisions on optimization and deoptimization. It is like a global Boolean flag. The language developer can invalidate an assumption programmatically to let the Truffle runtime know that a particular assumption is no longer valid, and accordingly the Truffle runtime can take decisions. Assumption objects are typically stored in the nodes as final fields. An assumption guard is used to pick a specialization method if the assumptions are true.

Based on these various annotations, Truffle will generate the actual `execute()` method with all the `if/else` controls to make sure the right version of the method is called based on the constraints that we declared with the `@Specification` annotation. The Truffle DSL annotation generator also includes `CompilerDirectives.transferToInterpreterAndInvalidate()` in the `execute()` method, toward the end; this will tell the compiler to stop the compilation, insert a transfer to the interpreter, and invalidate the machine code. This will trigger the deoptimization and a return to the interpreter mode of execution.

Apart from this, Truffle DSL also provides other annotations that make the job of a language developer easy. You can refer to the full list here: `https://www.graalvm.org/truffle/javadoc/com/oracle/truffle/api/dsl/package-summary.html`.

Truffle defines a TypeSystem, where the language developers can provide custom behavior for casting operand types. For example, the Truffle interpreter may not know how to typecast a `long` to `int`. Using TypeSystem, we can define the typecasting logic. The Truffle interpreter will use the TypeSystem during specialization.

One of the challenges with dynamically typed language is the polymorphic dispatch of the methods/functions. Truffle interpreters implement polymorphic inline caching to speed up the function lookups.

Polymorphic inline caching

In dynamically typed languages, the interpreters have to perform lookups to identify the right implementation of the method/function that is called. Looking up the functions and calling the functions is expensive and slows down the execution. In dynamically typed languages, when an object or function is called, the class is not declared, at build time or at run time, the interpreter has to do a lookup to find the actual class that is implementing the method. This is typically a hashtable lookup, unlike the vTable lookup that occurs in strongly typed languages. Hashtable lookups are time-consuming and very expensive and slow down execution. If we have only one class implementing the method, we have to only do the lookup once. This is called monomorphic inlining. If multiple classes implement the method, it's polymorphic.

Checking if the function lookup is valid is less expensive than the actual lookup. Truffle caches polymorphic lookups if there are lots of previous lookups for multiple (polymorphic) lookups of a function. When a function is redefined due to deoptimization, the `Assumption` object is used to invalidate and perform a fresh lookup. To improve the performance of lookups, Truffle provides the polymorphic inline cache. Truffle caches the lookups, and just checks if the lookup is still valid.

Understanding how Truffle supports interoperability

Truffle provides a very well-designed interoperability framework to allow guest languages to read and store data. In this section, we will cover some of the key features that the Truffle interoperability framework provides. Let's have a rundown of each of them.

Frame management and local variables

Truffle provides a standard interface to handle the local variables and data between host and guest language implementations. The frame provides the interface to read and store the data in the current namespace. When a function is called, the local variables' data is passed as an instance of `com.oracle.truffle.api.frame.Frame`. There are two implementations of the frame:

- **VirtualFrame**: This is most commonly used, and is passed as a parameter to the `execute()` method. This is lightweight, and preferable, as Graal optimizes this better. This frame lives in the scope of the function. It is the optimum and recommended way to pass data to functions. `VirtualFrame` does not escape, so it is easy to handle and inline.

- **MaterializedFrame**: `MaterializedFrame` is allocated in the heap and is accessible to other functions. `MaterializedFrame` lives beyond the scope of the function. Graal cannot optimize it as it might optimize `VirtualFrame`. This frame implementation also has an effect on the memory and speed.

Frames keep track of the type of data that is stored as part of the key. The key used to get the data is an instance of `FrameSlot` and `FrameSlotKind`. The following code snippet shows the `Frame` interface definition:

```
public interface Frame {
    FrameDescriptor getFrameDescriptor();
    Object[] getArguments();
    boolean isType(FrameSlot slot);
    Type getType(FrameSlot slot) throws
            FrameSlotTypeException;
    void setType(FrameSlot slot, Type value);
    Object getValue(FrameSlot slot);
    MaterializedFrame materialize();
}
```

`FrameSlot.getIdentifier()` provides the unique identifier for the data, and `FrameSlotKind` stores the type of data. `FrameSlotKind` is an enum of various types (Boolean, Byte, Double, Float, Illegal, Int , Long, Object).

The `FrameDescriptor` class keeps track of values stored in the frames. `FrameDiscriptor` describes the layout of `Frame`, providing a mapping of `FrameSlot` and `FrameSlotKind` and the value. Please refer to `https://www.graalvm.org/truffle/javadoc/com/oracle/truffle/api/frame/package-summary.html` for more details on the Frame API. SimpleLanguage has an implementation of frame management and is a good start to understand how the Frame API can be used to manage the data that is passed between languages while invoking methods/functions.

Dynamic Object Model

Truffle provides a **Dynamic Object Model (DOM)** that provides an object storage framework to enable the interoperability of data and objects between different languages. Truffle's DOM defines a standard and optimized way to share data, especially between dynamically typed languages. The DOM provides a language-independent shared infrastructure that allows developers to derive and implement various language implementations of the objects. This also helps us to share type objects between languages. Truffle's DOM is one of the core components of Truffle's interoperability and embedding feature. It provides a consistent in-memory object storage structure for host and guest languages. This allows the sharing of data between code written in different languages and the application of optimization across polyglot applications.

One of the challenges with dynamically typed languages is the dynamism that is expected of the data object model. The structure of the object may change dynamically. To support this, Truffle's DOM defines a Java class called `DynamicObject`. It provides extension arrays to provide the variability of primitive types and object extensions.

The guest language objects should all derive from a base class that extends from `DynamicObject` and implements `TruffleObject`. Let's now understand the Truffle instrumentation in detail.

Understanding Truffle instrumentation

Truffle provides an Instrumentation API to help build instrumentation and tools for diagnosis, monitoring, and debugging. Truffle also provides a reference implementation called Simple Tool (`https://github.com/graalvm/simpletool`). Truffle provides a very high-performance instrumentation design. The instrumentation is achieved with the help of probes and tags. The probes are attached to the AST nodes to capture the instrumentation data, and the nodes are identified using tags. Multiple instruments can be attached to the probe. The following figure shows a typical instrumentation:

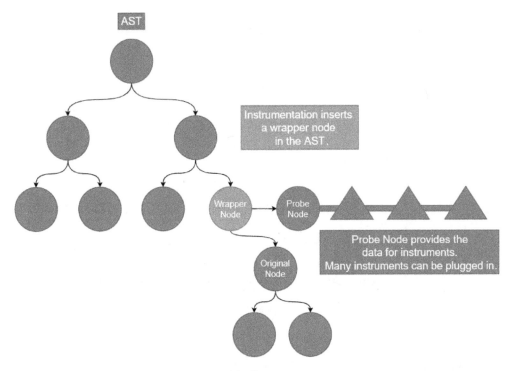

Figure 6.6 – Truffle instrumentation

The preceding figure illustrates how Truffle's Instrument API connects to the AST to collect the various metrics/data. Truffle displaces the original node by inserting a wrapper node and passes the information to the probe node, which can be connected to multiple instruments to collect the data.

Ahead-of-time compilation using Truffle

Guest language developers can make use of Graal's **Ahead-of-Time (AOT)** feature by implementing Truffle AOT. The guest language developer has to implement the `RootNode.prepareForAOT()` method by returning a non-null value. If a null value is returned, Truffle/Graal understands that this language does not support building native images. To support AOT, the `prepraeForAOT()` method typically might implement the following tasks:

- Provide type information about the local variables and update them in `FrameDescriptor`. This will help the AOT compiler to resolve the types during build time.

- Resolve and define the arguments and return types.

Truffle DSL provides helper classes to accelerate the development of AOT features. The `com.oracle.truffle.api.dsl.AOTSupport` class recursively prepares the AST for AOT. Each node in the AST has to have an implementation of the `prepareForAOT()` method.

AOT compilation can be triggered by passing the `--engine.CompileAOTOnCreate=true` argument to the language launcher. Each guest language will have a language launcher to run the application, for example, `js` for JavaScript, `graalpython` for Python, and so on. We will cover some of these guest language implementations in the next two chapters.

Optimizing Truffle interpreter performance with launcher options

Truffle defines a design and specification for providing various launcher options that can be used to diagnose, debug, and optimize the interpreter. All the guest language developers support these launcher options. In this section, we will cover some of these important launcher options:

- **Help**: All language launchers implement the `-help` command-line argument. `--help:expert` provides the expert options. For the language implementer's internal options, we can use `--help:internal`.

- **Analysis and profiling options**: Truffle provides command-line options to generate the Graal dumps that can be analyzed using the **Ideal Graph Visualizer**. Please refer to the *Installing the Ideal Graph Visualizer* and *Graal intermediate representation* sections in *Chapter 4, Graal Just-In-Time Compiler*, for more details on how to use Graal graphs for analysis and diagnosis. Graal graphs can be generated by passing the `--vm.Dgraal.Dump=Truffle:1` argument. The Graal graphs generated with Truffle will have a phase called *After TruffleTier* that shows the optimizations performed by Truffle.

  `--cpusampler` can be used to find out the CPU time taken to run the application and provide a detailed breakdown of CPU usage by module.

  The `--engine.TraceCompilation` argument can be passed to create a trace each time a method is compiled.

  The `--engine.TraceCompilationDetail` argument can be passed to trace when compilation is queued, started, and finished.

  The `--engine.TraceCompilationAST` argument can be passed to trace the AST whenever the code is compiled.

The `--engine.TraceInlining` argument can be passed to trace the inlining decisions taken by the guest language.

The `--engine.TraceSplitting` argument can be passed to trace the splitting decisions taken by the language.

The `--engine.TraceTransferToInterpreter` argument can be passed to trace when the deoptimization is triggered and a transfer to the interpreter occurs.

You can refer to the GraalVM documentation for more information (`https://www.graalvm.org/graalvm-as-a-platform/language-implementation-framework/Optimizing/`) or pass the `--help` argument in the language launcher.

SimpleLanguage and Simple Tool

The GraalVM team has created a reference implementation of a guest language called SimpleLanguage. SimpleLanguage demonstrates the features of Truffle and explains how to use the Truffle API. A guest language developer can use SimpleLanguage as a reference. It is completely open source and is available on GitHub at `https://github.com/graalvm/simplelanguage`. SimpleLanguage is just a starting point and does not implement all the features.

There is also a reference implementation of Simple Tool. Simple Tool is an implementation of a code coverage tool that has been built using Truffle. This is also an open source project that can be used by tool developers to build new tools using Truffle to run on GraalVM. You can access the source code of this tool at `https://github.com/graalvm/simpletool`.

There is an increasing number of languages being developed with Truffle. We will be covering JavaScript, LLVM (C/C++), Ruby, Python, R, Java/Truffle, and WebAssemby in the next two chapters. You can see the status of some of the other programming languages at `https://www.graalvm.org/graalvm-as-a-platform/language-implementation-framework/Languages/`.

Summary

In this chapter, we went through the architecture of Truffle and saw how it provides a well-designed framework for other languages (guest languages) to run on GraalVM. We also looked at how Truffle interpreters are implemented and how they can optimize the AST before submitting the stabilized AST to the Graal for further optimization.

In this chapter, you have gained a good understanding of Truffle architecture and how Truffle provides a framework and an implementation layer on top of Graal. You have also explored the optimizations Truffle performs before submitting the code to Graal JIT for further optimization and execution.

In the next chapter, we will look at how JavaScript and LLVM languages (C, C++, and so on) implement Truffle and run on GraalVM.

Questions

1. What is specialization?

2. What is tree/node rewriting?

3. What is partial evaluation?

4. What is Truffle DSL?

5. What is a frame?

6. What is a Dynamic Object Model?

Further reading

- Truffle: A Self-Optimizing Runtime System (`https://lafo.ssw.uni-linz.ac.at/pub/papers/2012_SPLASH_Truffle.pdf`)

- Specializing Dynamic Techniques For Implementing the Ruby Programming Language (`https://www.researchgate.net/publication/285051808_Specialising_Dynamic_Techniques_For_Implementing_the_Ruby_Programming_Language`)

- A Domain-Specific Language for Building Self-Optimizing AST Interpreters (`http://lafo.ssw.uni-linz.ac.at/papers/2014_GPCE_TruffleDSL.pdf`)

- One VM to Rule Them All (`http://lafo.ssw.uni-linz.ac.at/papers/2013_Onward_OneVMToRuleThemAll.pdf`)

- High-Performance Cross-Language Interoperability in a Multi-language Runtime (`https://chrisseaton.com/rubytruffle/dls15-interop/dls15-interop.pdf`)

- Writing a Language in Truffle (`http://cesquivias.github.io/index.html`)

7
GraalVM Polyglot – JavaScript and Node.js

In the previous chapter, we looked at how Truffle provides a layer to integrate other language programs to run on GraalVM. In this chapter, we will focus on JavaScript and Node.js interpreters, and in the next chapter, we will cover other runtimes, such as Java/Truffle, Python, R, and WebAssembly. We will be covering aspects of the polyglot interoperability features of Truffle and exploring the JavaScript interpreter. We will be exploring these features hands-on by writing code.

In this chapter, we will cover the following topics:

- Understanding how to run non-JVM language applications on Graal, specifically JavaScript and Node

- Learning how to pass objects/values between applications written in different languages

- Understanding how to use optimization techniques to fine-tune the code

By the end of this chapter, you will have a very clear understanding of how to build a polyglot application on GraalVM.

Technical requirements

In this chapter, we will be doing a lot of hands-on coding, to explore the various guest languages that GraalVM supports. To try the code, you will need the following:

- Various language Graal runtimes: We will cover in the chapter how to install and run these runtimes.

- Access to GitHub: There are some sample code snippets, which are available in a Git repository. The code can be downloaded from the following link. You will find the chapter-specific code under the `chapter7` directory: `https://github.com/PacktPublishing/Supercharge-Your-Applications-with-GraalVM/tree/main/Chapter07/js`.

- The Code in Action video for this chapter can be found at `https://bit.ly/3yqu4ui`.

Understanding the JavaScript (including Node.js) Truffle interpreter

The GraalVM version of JavaScript is an ECMAScript-compliant runtime for both JavaScript, `js`, and Node.js, `node`. It supports all the features of ECMAScript 2021, at the time of writing this book. It is also Nashorn- and Rhino-compatible and provides complete support for Node.js.

GraalVM Node.js uses the original Node.js source code and replaces the V8 JavaScript engine with the GraalVM JavaScript engine. The replacement is seamless, and the application developers don't have to modify any significant amount of code or configuration to run existing Node.js applications with GraalVM Node.js. GraalVM Node. js provides more features for embedding code from other languages, and accessing data and code, and interoperating code in other language. **Node package manager** (**NPM**) is also included in npm.

In this section, apart from using JavaScript and Node as an alternate runtime for running applications, we will also explore their polyglot interoperability features. We will be walking through a lot of JavaScript and Node.js sample code to explore the polyglot capabilities of the GraalVM JavaScript runtime.

Verifying JavaScript, Node, and npm installation and versions

JavaScript and Node.js come along with the GraalVM installation; you will find them in the `<GraalHome>/bin` directory. We can check if the JavaScript runtime is properly configured by checking the version number.

To check the version, execute the `js --version` command. At the time of writing this book, GraalVM JavaScript 21.0.0.2 was the latest. The following is the output (note that it is GraalVM JavaScript):

```
GraalVM JavaScript (GraalVM EE Native 21.0.0.2)
```

We can also ensure that we are running the right version of Node.js by calling a specific node.js version by executing the `node --version` command. In the following command, we are explicitly calling the right version. Note that the GraalVM home location might be different for you:

```
/Library/Java/JavaVirtualMachines/graalvm-ee-java11-21.0.0.2/
Contents/Home/bin/node --version
v12.20.1
```

Let's also make sure NPM is working by executing the `npm --version` command. The following is the command and output:

```
/Library/Java/JavaVirtualMachines/graalvm-ee-java11-21.0.0.2/
Contents/Home/bin/npm --version
6.14.10
```

Now that we have verified the JavaScript, Node.js, and npm installations, let's create a simple Node.js application.

Go to an application folder, and execute `npm init`. This will set up the boilerplate configuration for the Node.js application. We will name the application `graal-node-app`. The following shows the console output:

```
/Library/Java/JavaVirtualMachines/graalvm-ee-java11-21.0.0.2/
Contents/Home/bin/npm init
package name: (npm) graal-node-app
version: (1.0.0) 1.0.0
description:
entry point: (index.js)
test command:
```

```
git repository:
keywords:
author:
license: (ISC)
About to write to /chapter7/js/npm/package.json:
{
  "name": "graal-node-app",
  "version": "1.0.0",
  "description": "",
  "main": "index.js",
  "scripts": {
    "test": "echo \"Error: no test specified\" andand exit 1"
  },
  "author": "",
  "license": "ISC"
}
Is this OK? (yes)
```

This creates a Node.js application and a `package.json` file with a boilerplate
configuration, based on the options we selected. Let's install the `express` package by
executing `npm install --save express`. This installs the `express` package in
the application folder and also updates the `package.json` file (because of the `--save`
argument). Here is the output:

```
/Library/Java/JavaVirtualMachines/graalvm-ee-java11-21.0.0.2/
Contents/Home/bin/npm install --save express
npm notice created a lockfile as package-lock.json. You should
commit this file.
npm WARN graal-node-app@1.0.0 No description
npm WARN graal-node-app@1.0.0 No repository field.

+ express@4.17.1
added 50 packages from 37 contributors and audited 50 packages
in 6.277s
found 0 vulnerabilities
```

You will find the node_modules directory, which contains all the packages that are required to run our application. Let's create an index.js file with the following code:

```
var express = require('express');
var app = express();

app.get('/', function(request, response) {
    var responseString = "<h1>Hello Graal Node </h1>";
    response.send(responseString);
});

app.listen(8080, function() {
    console.log('Started the server at 8080')
});
```

As you can see, it's a very simple application, which responds with an HTML Hello Graal Node as a header 1, when invoked at the root. The application will listen at port number 8080.

Let's run this application with the following command:

```
/Library/Java/JavaVirtualMachines/graalvm-ee-java11-21.0.0.2/
Contents/Home/bin/node index.js
Started the server at 8080
```

Now that we can see the output, we know that the application is listening at 8080. Let's try to call this from the web browser at http://localhost:8080/. The following is the screenshot of the application response on the web browser:

Figure 7.1 – Hello Graal Node screenshot

Now that we know that the Node.js on GraalVM is working fine, let's understand the polyglot interoperability.

JavaScript interoperability

In *Chapter 6, Truffle for Multi-language (Polyglot) support*, we went into a lot of details about how Truffle enables polyglot support and provides an infrastructure for polyglot interoperability and polyglot embedding. In this section, we will explore these features with sample code.

Let's take the Node.js application that we created in the previous section, and add an endpoint, /poly, in our index.js file. Let's create a simple Python array object, and store some numbers in the object. We will then iterate through this Python object in Node.js, to list these numbers. This shows how we can embed Python code snippets within JavaScript.

The following source code shows this new endpoint, /poly:

```
var express = require('express');
var app = express();
app.get('/', function(request, response) {
    var responseString = "<h1>Hello Graal Node </h1>";
    response.send(responseString);
});
app.get('/poly', function(request, response) {

    var responseString = "<h1>Hello Graal Polyglot </h1>";
    var array = Polyglot.eval("python", "[1,2,3,4, 100,
                              200, 300, 400]")
    responseString = responseString + "<ul>";
    for (let index = 0; index < array.length; index++) {
        responseString = responseString + "<li>";
        responseString = responseString + array[index];
        responseString = responseString + "</li>";
    }
    responseString = responseString + "</ul>";
    response.send(responseString);
});
```

Now let's have the code to listen at port 8080 and call the preceding function when we receive a request:

```
app.listen(8080, function() {
    console.log('Started the server at 8080')
});
```

As you can see in the code, we are using the `Polyglot.eval()` method to run Python code. To let the polyglot object know that it's Python code, we are passing `python` as a parameter, and passing the Python representation of the array. Let's now run this code with node:

```
/Library/Java/JavaVirtualMachines/graalvm-ee-java11-21.0.0.2/
Contents/Home/bin/node --jvm --polyglot index.js
Started the server at 8080
```

Note that we have to pass `--jvm` and `--polyglot` arguments to node. It's very important to pass these parameters. `--jvm` tells node to run on **Java Virtual Machine (JVM)**, and `--polyglot`, as the name suggests, tells node to support `polyglot`. Since Truffle and Graal run on JVM, it's important to use the `jvm` argument, even though we may not be directly using Java in our code.

Let's now access this new endpoint from the browser. The following screenshot shows the output as expected:

Hello Graal Polyglot

- 1
- 2
- 3
- 4
- 100
- 200
- 300
- 400

Figure 7.2 – /poly endpoint result screenshot

As you might have noticed, the first time we called, it took time to load the page, but the subsequent calls are instantaneous. Let's just time that with curl (curl is a command-line utility to call any URLs. Please refer to `https://curl.se/` for more details on curl and how to install curl on your machine). The following is a screenshot of a sequence of curl commands:

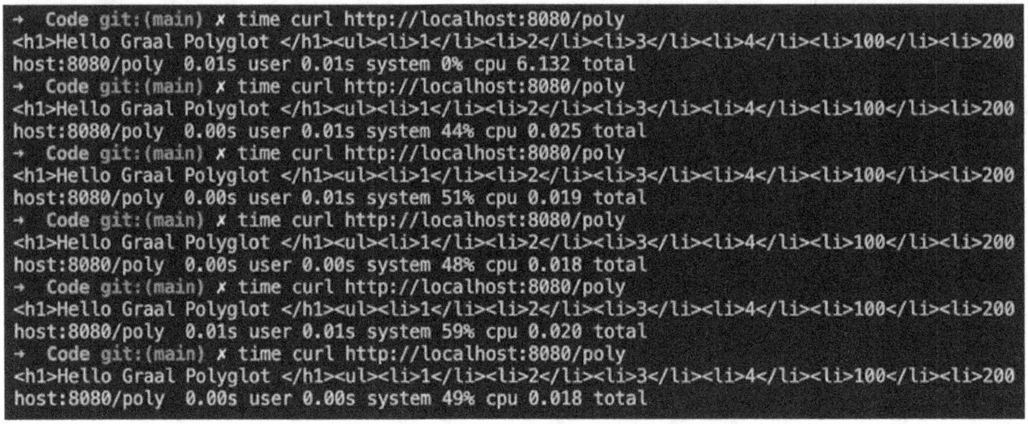

```
→  Code git:(main) x time curl http://localhost:8080/poly
<h1>Hello Graal Polyglot </h1><ul><li>1</li><li>2</li><li>3</li><li>4</li><li>100</li><li>200
host:8080/poly  0.01s user 0.01s system 0% cpu 6.132 total
→  Code git:(main) x time curl http://localhost:8080/poly
<h1>Hello Graal Polyglot </h1><ul><li>1</li><li>2</li><li>3</li><li>4</li><li>100</li><li>200
host:8080/poly  0.00s user 0.01s system 44% cpu 0.025 total
→  Code git:(main) x time curl http://localhost:8080/poly
<h1>Hello Graal Polyglot </h1><ul><li>1</li><li>2</li><li>3</li><li>4</li><li>100</li><li>200
host:8080/poly  0.00s user 0.01s system 51% cpu 0.019 total
→  Code git:(main) x time curl http://localhost:8080/poly
<h1>Hello Graal Polyglot </h1><ul><li>1</li><li>2</li><li>3</li><li>4</li><li>100</li><li>200
host:8080/poly  0.00s user 0.00s system 48% cpu 0.018 total
→  Code git:(main) x time curl http://localhost:8080/poly
<h1>Hello Graal Polyglot </h1><ul><li>1</li><li>2</li><li>3</li><li>4</li><li>100</li><li>200
host:8080/poly  0.01s user 0.01s system 59% cpu 0.020 total
→  Code git:(main) x time curl http://localhost:8080/poly
<h1>Hello Graal Polyglot </h1><ul><li>1</li><li>2</li><li>3</li><li>4</li><li>100</li><li>200
host:8080/poly  0.00s user 0.00s system 49% cpu 0.018 total
```

Figure 7.3 – Performance of Node.js after subsequent calls

We can see the initial load on the CPU, but subsequent calls are quick with no additional load on the CPU.

Let's now explore more advanced features of polyglot interoperability. JavaScript and Java interoperability is very sophisticated. Let's explore the concepts with more complex implementations than lists.

JavaScript embedded code in Java

Let's recall the `FibonaaciCalculator.java` file that we used in previous chapters. Let's modify that `FibonacciCalculator.java` to use a JavaScript snippet and execute that JavaScript snippet within Java.

Here is the modified version of the `FibonacciCalculator` with an embedded JavaScript snippet. The Java file is called `FibonacciCalculatorPolyglot.java`. You can find the full code in the Git repository:

```
import org.graalvm.polyglot.*;
import org.graalvm.polyglot.proxy.*;
```

We have to import the `polyglot` classes. This implements the Truffle interoperability:

```java
public class FibonacciCalculatorPolyglot{
    static String JS_SNIPPET =
    "(function logTotalTime(param){console.log('total(from
        JS) : '+param);})";
    public int[] findFibonacci(int count) {
        int fib1 = 0;
        int fib2 = 1;
        int currentFib, index;
        int [] fibNumbersArray = new int[count];
        for(index=2; index < count; ++index ) {
            currentFib = fib1 + fib2;
            fib1 = fib2;
            fib2 = currentFib;
            fibNumbersArray[index - 1] = currentFib;
        }
        return fibNumbersArray;
    }
```

Now let's define the `main()` function, which will invoke `findFibonacci()` several times to reach the compiler threshold:

```java
public static void main(String args[]){
    FibonacciCalculatorPolyglot fibCal
        = new FibonacciCalculatorPolyglot();
    long startTime = System.currentTimeMillis();
    long now = 0;
    long last = startTime;
    for (int i = 1000000000; i < 1000000010; i++) {
        int[] fibs = fibCal.findFibonacci(i);

        long total = 0;
        for (int j=0; j<fibs.length; j++) {
            total += fibs[j];
        }
        now = System.currentTimeMillis();
        System.out.printf("%d (%d ms)%n", i , now - last);
```

```
        last = now;
    }
    long endTime = System.currentTimeMillis();
    long totalTime =
        System.currentTimeMillis() - startTime;
    System.out.printf("total (from Java): (%d ms)%n",
                      totalTime);
    try (Context context = Context.create()) {
        Value value = context.eval("js", JS_SNIPPET);
        value.execute(totalTime);
    }
    }
}
```

Let's explore this code. We have defined a static `String` variable that holds the JavaScript snippet, shown next:

```
static String JS_SNIPPET = "(function logTotalTime(param)
{console.log('total(from JS) : '+param);})";
```

We have defined a static `String` with a simple JavaScript function that prints whatever is the parameter that is passed to it. To invoke this JavaScript code within Java, we need to first import the `Polyglot` libraries by importing the following packages:

```
import org.graalvm.polyglot.Context;
import org.graalvm.polyglot.Value;
```

To invoke the JavaScript code, we first need to create an instance of the `org.graalvm.polyglot.Context` class. The `Context` object provides the polyglot context to allow the guest language code to run in the host language. A polyglot context represents the global runtime state of all installed and permitted languages.

The simplest way to use the `Context` object is to create the `Context` object and use the `eval()` function in the `Context` object, to execute other language code. The following is a code snippet where we are executing a JavaScript code snippet within Java. In this case, the guest language is JavaScript, which is passed as a parameter, `"js"`, in the `eval` method in the host language, Java:

```
try (Context context = Context.create()) {
    Value value = context.eval("js", JS_SNIPPET);
```

```
    value.execute(totalTime);
}
```

Now let's execute this code. Following is the screenshot of the output, after execution:

```
→ npm git:(main) x java FibonacciCalculatorPolyglot
1000000000 (3754 ms)
1000000001 (671 ms)
1000000002 (1115 ms)
1000000003 (845 ms)
1000000004 (882 ms)
1000000005 (891 ms)
1000000006 (871 ms)
1000000007 (842 ms)
1000000008 (865 ms)
1000000009 (887 ms)
total (from Java): (11623 ms)
total(from JS) : 11623
```

Figure 7.4 – Output of FibonacciCalculatorPolyglot showing Java and JavaScript outputs

As you can see in the output, we have two totals printed, one is printed with Java code and the other one is printed from the JavaScript code.

This opens up a lot of possibilities – imagine running machine learning code written in Python or R in Node.js web applications. We are bringing the best features of individual languages together in one VM.

The Context object has ContextBuilder, which can be used to set specific environment properties. The following are some of the properties that it can set, and the relevant Context creates code. This can be used to control the access the guest language has to the host. The code to control the access is Context.newBuilder(). allowXXX().build(). The following are various allowXXX methods that can be used for finer access control:

- allowAllAccess(boolean): This is the default. It provides all access to the guest language.

- allowCreateProcess(boolean): Provides control access for the guest language to create a new process.

- allowCreateThread(boolean): Provides control access for the guest language to create a new thread.

- allowEnvironmentAccess(EnvironmentAccess): Allows control access to the environment using the provided policy.

- `allowHostClassLoading(boolean)`: This allows the guest languages to load new host classes via a JAR or a class file.

- `allowIO(boolean)`: Controls access to perform I/O operations. If true, the guest language can perform unrestricted I/O operations on the host system.

- `allowNativeAccess(boolean)`: Controls guest languages to access the native interface.

- `allowPolyglotAccess(PolyglotAccess)`: Controls polyglot access using the provided policy. `PolyglotAccess` can be used to define custom polyglot access policies on how the data, bindings, and code execution can be the controlled at a finer level. This is a custom implementation, which the guest languages can build using the `PolyglotAccess` builder.

Refer to the Javadoc (`https://www.graalvm.org/truffle/javadoc/org/graalvm/polyglot/Context.html`) for more details about other methods. It is risky to give the guest language all the access; it's always better to provide fine and specific access based on the requirement.

Here is an example of how we can build a `Context` object with specific access:

```
Context context = Context.newBuilder().allowIO(true).build();
```

We can also load an external file using the following code snippet, which is a recommended way of embedding code. It is not a good practice to copy-paste the other language code as a string into the host language. It's a configuration management nightmare to keep the code up to date and bug-free, as the code in other languages might be developed by different developers.

Following is a code snippet that shows how to load the source code as a file, rather than embedding the guest language code in the host source code:

```
Context ctx =
    Context.newBuilder().allowAllAccess(true).build();
    File path = new File("/path/to/scriptfile");
    Source pythonScript =
        Source.newBuilder("python", new File(path,
            "pythonScript.py")).build();
    ctx.eval(pythonScript)
```

In this section, we saw how we can invoke JavaScript code from Java. Now let's try to call a Java class from JavaScript.

Calling a Java class from JavaScript/Node.js

Now that we have seen how Java code can run JavaScript code, let's try to call Java code from JavaScript. Here is a very simple Java application, which prints the argument that is passed to it on a console. The name of the Java file is `HelloGraalPolyglot.java`:

```java
public class HelloGraalPolyglot {
    public static void main(String[] args) {
        System.out.println(args[0]);
    }
}
```

Let's compile this application with `javac HelloGraalPolyglot.java`.

Now let's try to call this application from the JavaScript. The following is the JavaScript code `hellograalpolyglot.js`:

```javascript
var hello = Java.type('HelloGraalPolyglot');
hello.main(["Hello from JavaScript"]);
```

It is very simple JavaScript code. We are loading the Java class using the `Java.type()` method in JavaScript, and calling the `main()` method with a `String` parameter, and passing the string `"Hello from JavaScript"`.

To execute this JavaScript, we will have to pass the `--jvm` argument and `--vm.cp` to set the classpath. Here is the command:

```
js --jvm --vm.cp=. hellograalpolyglot.js
```

The following shows the output of executing this command:

```
js --jvm --vm.cp=. hellograalpolyglot.js
Hello from JavaScript
```

This was a very simple example. To understand how parameters are passed and how the method return data is captured and used in JavaScript, let's try to call the `findFibonacci()` method defined in the `FibonacciCalculator.java` code, from a Node.js application. We will pass a parameter and get an array out of the method, which we will render as a web page.

Let's modify `index.js` and add another endpoint, `/fibonacci`. Here is the complete source code:

```
app.get('/fibonacci', function(request, response) {
    var fibonacciCalculatorClass =
        Java.type("FibonacciCalculatorPolyglot");
    var fibonacciCalculatorObject = new
        fibonacciCalculatorClass();
    //fibonacciCalculatorClass.class.static.main([""]);
    var array =
        fibonacciCalculatorObject.findFibonacci(10);
    var responseString =
        "<h1>Hello Graal Polyglot - Fibonacci numbers </h1>";
    responseString = responseString + "<ul>";
    for (let index = 0; index < array.length; index++) {
        responseString = responseString + "<li>";
        responseString = responseString + array[index];
        responseString = responseString + "</li>";
    }
    responseString = responseString + "</ul>";
    response.send(responseString);
});
```

In this `node.js` code, we are first loading the Java class `FibonacciCalculatorPolyglot` using the `Java.Type()` method. Then we are creating an instance of this class and calling the method directly. The output, we know, is an array. We are iterating through the array and printing the result as an HTML list.

Let's run this code with the following command:

```
/Library/Java/JavaVirtualMachines/graalvm-ee-java11-21.0.0.2/
Contents/Home/bin/node --jvm --polyglot index.js
Started the server at 8080
```

Now let's go to `http://localhost:8080/fibonacci`. Here is the screenshot of the output:

Hello Graal Polyglot - Fibonacci numbers

- 0
- 1
- 2
- 3
- 5
- 8
- 13
- 21
- 34
- 0

Figure 7.5 – Output of the Node.js application calling the FibonacciCalculator method screenshot

The preceding screenshot shows the Node.js/Fibonacci endpoint working, where it is listing the first 10 Fibonacci numbers as an HTML list.

In this section, we looked at how to run a JavaScript snippet within Java, invoke Java from JavaScript and invoke a Java method, pass parameters, and get results from a Java method from a Node.js application. Let's very quickly summarize the various JavaScript interoperability features:

- When we want to call Java code from JavaScript, we pass the `--jvm` argument and set `CLASSPATH` to load the right class using `--vm.cp`.

- We use the polyglot `Context` object in Java to run other language code. There is a special `ScriptEngine` object for running JavaScript in Java. The `Context` object wraps this and is the recommended way to run.

- We use `Java.type()` to load a Java class from JavaScript/Node.js.

- We can use `new` to create the instances of the class.

- Type conversion is taken care of by GraalVM between Java and JavaScript. In cases where there could be a loss of data (for example, converting from `long` to `int`) a `TypeError` is thrown.

- Java package resolution can be done by providing the full package path while calling `Java.type()`.

- Exception handling can be done naturally using `try{}catch` blocks both in Java and JavaScript. GraalVM takes care of converting the exception.

- In the preceding example, we looked at how Java arrays can be iterated by JavaScript. Similarly, `Hashmap` can also be used natively using the `put()` and `get()` methods.

- JavaScript objects can be accessed by Java code as instances of the `com.oracle.truffle.api.interop.java.TruffleMap` class.

In this section, we looked at how we can interoperate between Java and JavaScript. Let's now explore how to build polyglot native images.

Polyglot native images

Graal also supports creating native images of polyglot applications. To create a native image of this Java class, we have to use the `--language` argument to build the native image. The following are the various language flags we can pass to `native-image` (the Native Image builder). In *Chapter 5*, *Graal Ahead-of-Time Compiler and Native Image*, we covered the Native Image builder in detail:

```
--language:nfi
--language:python
--language:regex
--language:wasm
--language:java
--language:llvm
--language:js
--language:ruby
```

In our example, we have to pass `--language:js` to let the Native Image builder know that we are using JavaScript within our Java code. So, we need to execute the following command:

```
native-image --language:js FibonacciCalculatorPolyglot
```

Following is the screenshot of the output after executing the command:

```
→  npm git:(main) x native-image --language:js FibonacciCalculatorPolyglot
[fibonaccicalculatorpolyglot:52735]        classlist:    1,294.00 ms,   0.96 GB
[fibonaccicalculatorpolyglot:52735]           (cap):    7,335.44 ms,   0.96 GB
[fibonaccicalculatorpolyglot:52735]          setup:    9,115.33 ms,   0.96 GB
[fibonaccicalculatorpolyglot:52735]        (clinit):    1,330.23 ms,   6.29 GB
[fibonaccicalculatorpolyglot:52735]      (typeflow):   20,887.35 ms,   6.29 GB
[fibonaccicalculatorpolyglot:52735]       (objects):   14,137.51 ms,   6.29 GB
[fibonaccicalculatorpolyglot:52735]      (features):    5,000.06 ms,   6.29 GB
[fibonaccicalculatorpolyglot:52735]       analysis:   43,815.99 ms,   6.29 GB
[fibonaccicalculatorpolyglot:52735]       universe:    1,593.65 ms,   6.29 GB
10115 method(s) included for runtime compilation
[fibonaccicalculatorpolyglot:52735]         (parse):    5,726.79 ms,   6.21 GB
[fibonaccicalculatorpolyglot:52735]        (inline):    4,240.51 ms,   5.88 GB
[fibonaccicalculatorpolyglot:52735]       (compile):   62,526.28 ms,   7.68 GB
[fibonaccicalculatorpolyglot:52735]        compile:   77,927.23 ms,   7.63 GB
[fibonaccicalculatorpolyglot:52735]          image:   11,543.48 ms,   7.63 GB
[fibonaccicalculatorpolyglot:52735]          write:    2,626.77 ms,   7.49 GB
[fibonaccicalculatorpolyglot:52735]        [total]:  149,737.75 ms,   7.49 GB
```

Figure 7.6 – Polyglot Native Image build output screenshot

The Native Image builder performs a static code analysis and builds the optimum image of our polyglot application. We should be able to find the executable `fibonaccicalculatorpolyglot` file in the directory. Let's execute the native image with the following command:

```
./fibonaccicalculatorpolyglot
```

The following figure shows the screenshot of the output when we run the native image:

```
→  npm git:(main) x ./fibonaccicalculatorpolyglot
1000000000 (2490 ms)
1000000001 (2758 ms)
1000000002 (2646 ms)
1000000003 (2679 ms)
1000000004 (2672 ms)
1000000005 (2636 ms)
1000000006 (2603 ms)
1000000007 (2631 ms)
1000000008 (2628 ms)
1000000009 (3232 ms)
total (from Java): (26975 ms)
total(from JS) : 26975
```

Figure 7.7 – Polyglot Native Image execution results screenshot

(In this example, you might find the code is performing more slowly than in JIT mode. Please refer to *Chapter 4, Graal Just-In-Time Compiler*, for more details on why this is happening.)

Bindings

The binding object acts as an intermediate layer between Java and JavaScript to access methods, variables, and objects between the languages. To understand how bindings work, let's write a very simple JavaScript file that has three methods – add(), subtract(), and multiply(). All three methods access two numbers and return a number. We also have a variable that holds a simple string. Here is the JavaScript code, Math.js:

```
var helloMathMessage = " Hello Math.js Variable";
function add(a, b) {
    return a+b;
}
function subtract(a, b) {
    return a-b;
}
function multiply(a, b) {
    return a*b;
}
```

This JavaScript code is very simple and straightforward.

Let's now write a simple Java class that loads this JavaScript file and calls the methods by passing integer parameters, and prints the result returned by JavaScript methods. This class also accesses the variable helloMathMessage and prints it.

Let's walk through the code to understand how this works. Here is the code, MathJSCaller.java:

```
import java.io.File;
import org.graalvm.polyglot.Context;
import org.graalvm.polyglot.Source;
import org.graalvm.polyglot.Value;
```

We are importing all the polyglot classes that implement the Truffle interoperability:

```
public void runMathJS() {
    Context ctx = Context.create("js");
    try {
        File mathJSFile = new File("./math.js");
        ctx.eval(Source.newBuilder("js", mathJSFile).build());
```

In the preceding code, we are creating the `Context` object and loading the JavaScript file and building it. Once the JavaScript is loaded, then to access the method members and variable member from the JavaScript file, we are using `Context.getBindings()`. Bindings provide a layer that allows polyglot languages to access the data and method members:

```
Value addFunction =
    ctx.getBindings("js").getMember("add");
Value subtractFunction =
    ctx.getBindings("js").getMember("subtract");
Value multiplyFunction =
     ctx.getBindings("js").getMember("multiply");
Value helloMathMessage =
    ctx.getBindings("js").getMember("helloMathMessage");
System.out.println("Binding Keys :" +
    ctx.getBindings("js").getMemberKeys());
```

We are just printing the binding keys to see what all members are exposed to. Now, let's access the members, by calling the methods and accessing the variable:

```
Integer addResult = addFunction.execute(30, 20).asInt();
Integer subtractResult = subtractFunction.execute(30,
        20).asInt();
Integer multiplyResult = multiplyFunction.execute(30,
        20).asInt();
System.out.println(("Add Result "+ addResult+ "
    Subtract Result "+ subtractResult+ " Multiply
    Result "+ multiplyResult));
System.out.println("helloMathMessage : " +
    helloMathMessage.toString());
}
```

Finally, we are printing all the results. The complete source code is available at the Git repository link provided in the *Technical requirements* section.

Now, let's run this application. The following screenshot shows the output:

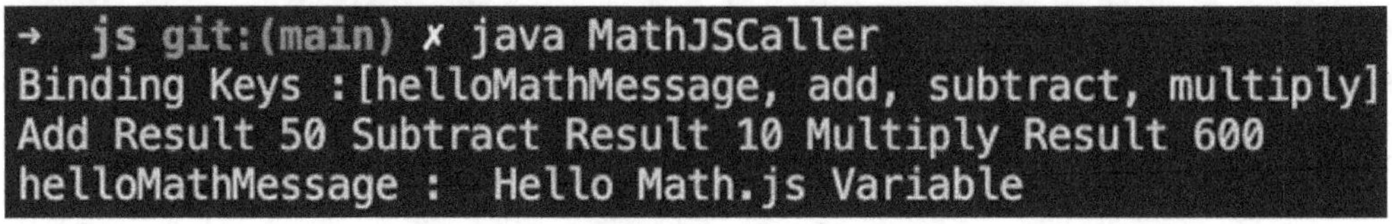

```
→  js git:(main) ✗ java MathJSCaller
Binding Keys :[helloMathMessage, add, subtract, multiply]
Add Result 50 Subtract Result 10 Multiply Result 600
helloMathMessage :  Hello Math.js Variable
```

Figure 7.8 MathJSCaller execution results

We can see that our program is working. It can load the JavaScript math.js file and call all the methods. We also see the list of binding keys, which we printed by calling System.out.println("Binding Keys :" + ctx.getBindings("js"). getMemberKeys());. We can see the list has four keys, and they match what we have in the math.js file.

In this example, we saw how a binding object acts as an interface to access JavaScript members from Java.

Multithreading

JavaScript on GraalVM supports multithreading. In this section, we will explore various patterns that are supported in the context of polyglot between Java and JavaScript.

A JavaScript object that is created in a thread can only be used within that thread, it cannot be accessed from another thread. For example, in our example, Value objects such as addFunction, subtractFunction, and so on can only be used with that thread.

Let's modify our MathJSCaller class' runMathJS() method to run a thread indefinitely, to simulate a concurrent access situation. Let's modify the preceding code and call the member functions in a separate thread. Here is the code snippet:

```java
Thread thread = new Thread(new Runnable() {
    @Override
    public void run() {
        while (true) {
            Integer addResult =
                addFunction.execute(30, 20).asInt();
            Integer subtractResult =
                subtractFunction.execute(30, 20).asInt();
            Integer multiplyResult =
                multiplyFunction.execute(30, 20).asInt();
```

```
            }
        }
    });
    thread.start();
```

We copied the accessing of the member methods in a separate thread. Now let's call this in a loop, to simulate concurrent access, using the same Context object within the threads and outside the thread. The following code snippet shows calls outside the thread using the same Context object:

```
while (true) {
    Integer addResult =
        addFunction.execute(30, 20).asInt();
    Integer subtractResult =
        subtractFunction.execute(30, 20).asInt();
    Integer multiplyResult =
        multiplyFunction.execute(30, 20).asInt();
    }
    } catch (Exception e) {
        System.out.println("Exception : " );
        e.printStackTrace();
    }
}
```

When we run this code, at some point, when the objects are simultaneously accessed by the two threads, we should get the following exception:

```
$ java MathJSCallerThreaded
(docker-desktop/bozo-book-library-dev)
Binding Keys :[helloMathMessage, add, subtract, multiply]
java.lang.IllegalStateException: Multi threaded access
requested by thread Thread[Thread-3,5,main] but is not allowed
for language(s) js.
…..
```

To overcome this issue, it is recommended to use isolated runtimes. We can create separate Context objects per thread and create new instances of these objects and use them in that thread. Here is the fixed code:

```
public void runMathJS() {
    Context ctx = Context.create("js");
    try {
        File mathJSFile = new File("./math.js");
        ctx.eval(Source.newBuilder
            ("js", mathJSFile).build());

        Value addFunction =
            ctx.getBindings("js").getMember("add");
        Value subtractFunction =
            ctx.getBindings("js").getMember("subtract");
        Value multiplyFunction =
            ctx.getBindings("js").getMember("multiply");
        Value helloMathMessage =
            ctx.getBindings("js")
            .getMember("helloMathMessage");

        System.out.println("Binding Keys :" + ctx.
        getBindings("js").getMemberKeys());
        while (true) {
            Integer addResult =
                addFunction.execute(30, 20).asInt();
            Integer subtractResult =
                subtractFunction.execute(30, 20).asInt();
            Integer multiplyResult =
                multiplyFunction.execute(30, 20).asInt();
}
```

Now, within the thread, we are creating a separate `Context` object. The following code snippet shows the updated code:

```
Thread thread = new Thread(new Runnable() {
    @Override
    public void run() {
        try {
            Context ctx = Context.create("js");
            ctx.eval(Source.newBuilder("js",
                mathJSFile).build());
            Value addFunction =
                ctx.getBindings("js").getMember("add");
            Value subtractFunction =
                ctx.getBindings("js").getMember("subtract");
            Value multiplyFunction =
                ctx.getBindings("js").getMember("multiply");
            Value helloMathMessage =
                ctx.getBindings("js")
                .getMember("helloMathMessage");
            while (true) {
                Integer addResult =
                addFunction.execute(30, 20).asInt();
                Integer subtractResult =
                    subtractFunction.execute(30, 20).asInt();
                Integer multiplyResult =
                    multiplyFunction.execute(30, 20).asInt();
            }
        } catch (Exception e) {
            e.printStackTrace();
        }
    }
});
thread.start();
```

As we can see, in this code, we are creating a separate `context` object within the thread, which is local to the thread. This does not create an exception.

The other solution to this is to access the `context` object in proper `synchronized` blocks or methods, so that the runtimes are not accessed at the same time. Here is the updated code, with a `synchronized` block:

```java
Thread thread = new Thread(new Runnable() {
    @Override
    public void run() {
        try {
            // Solution 2
            while (true) {
                synchronized(ctx) {
                    Integer addResult =
                        addFunction.execute(30, 20).asInt();
                    Integer subtractResult =
                        subtractFunction.execute(30, 20).asInt();
                    Integer multiplyResult =
                        multiplyFunction.execute(30, 20).asInt();
                }
            }
        } catch (Exception e) {
            e.printStackTrace();
        }
    }
});
thread.start();
```

We can also include the whole block as a synchronized block, still using the same `Context` object:

```java
while (true) {
    synchronized(ctx) {
        Integer addResult =
            addFunction.execute(30, 20).asInt();
        Integer subtractResult =
            subtractFunction.execute(30, 20).asInt();
        Integer multiplyResult =
            multiplyFunction.execute(30, 20).asInt();
    }
}
```

This will also run fine but might run slower than the previous solution, as there could be a lot of locks on the `Context` object.

Java objects are thread-safe, so Java objects can be accessed between JavaScript runtimes running different threads.

Asynchronous programming – Promise and await

Asynchronous programming is very prominent in modern distributed applications. JavaScript uses `Promise`. The `Promise` object represents the completion of an asynchronous activity, along with the final value. The `Promise` object has three states:

- **Pending**: This state is the initial state.
- **Fulfilled**: This state indicates that the operation successfully executed.
- **Rejected**: This state indicates that the operation failed.

Sometimes, we may have to have JavaScript creating a promise and the logic might be running in Java code, and when the Java code is done, it may have to fulfill or reject the promise. To handle that, Graal provides a `PromiseExecuter` interface. A Java class has to implement this interface method, `void onPromiseCreation(Value onResolve, Value onReject);`. A Java class that implements this interface can be used by JavaScript to create a `Promise` object. JavaScript can call `await` on a Java object that implements void then `(Value onResolve, Value onReject);` to implement asynchronous programming between JavaScript and Java.

Summary

In this chapter, we went through the various polyglot interoperability and embedding features of GraalVM/Truffle for JavaScript and Node.js in detail. We explored all the key concepts with some real code examples, to gain a clear understanding of how JavaScript and Node.js can call, pass data, and interoperate with other language code. This is one of the salient features of GraalVM.

The examples given in this chapter will help you to build and run polyglot applications that are written using the Java and JavaScript languages on the same runtime.

In the next chapter, we will continue to explore R, Python, and the latest Java on Truffle.

Questions

1. What JavaScript object and method is used to run other language code?

2. What is the `Context` object in Java?

3. How do you control the access a guest language gets to the host?

4. How do you build a native image of a polyglot application?

5. What is a binding?

Further reading

- GraalVM Enterprise Edition (`https://docs.oracle.com/en/graalvm/enterprise/19/index.html`)

- JavaScript and Node.js reference (`https://www.graalvm.org/reference-manual/js/`)

- *Truffle: A Self-Optimizing Runtime System* (`https://lafo.ssw.uni-linz.ac.at/pub/papers/2012_SPLASH_Truffle.pdf`)

- *An Object Storage Model for the Truffle Language Implementation Framework* (`https://chrisseaton.com/rubytruffle/pppj14-om/pppj14-om.pdf`)

8
GraalVM Polyglot – Java on Truffle, Python, and R

In the previous chapter, we covered JavaScript and Node.js interpreters and interoperability between languages. In this chapter, we will cover other language implementations such as the following:

- Java on Truffle (also called Espresso): Java implementation on Truffle

- GraalPython: Python language interpreter implementation

- FastR: R language interpreter implementation

All of these language implementations are still in the *experimental* phase so are not released for production at the time of writing the book. However, we will explore the features and build some code to understand the various concepts.

In this chapter, we will cover the following topics:

- Understanding Python, R, and Java/Truffle interpreters

- Learning about and exploring language interoperability

- Understanding the compatibility and limitations of these various language interpreters

By the end of this chapter, you will have hands-on experience in building polyglot applications with Python, R, and Java/Truffle interpreters.

Technical requirements

This chapter requires the following to follow along with the various coding/hands-on sections:

- The latest version of GraalVM.

- Various language Graal runtimes. We will cover in the chapter how to install and run these runtimes.

- Access to GitHub: There are some sample code snippets, which are available in a Git repository. The code can be downloaded from the following link: `https://github.com/PacktPublishing/Supercharge-Your-Applications-with-GraalVM/tree/main/Chapter08`.

- The Code in Action video for this chapter can be found at `https://bit.ly/3fj2iIr`.

Understanding Espresso (Java on Truffle)

GraalVM 21.0 is a major release that introduces a new guest language runtime called Java on Truffle. Before this, we had the option to run Java using HotSpot (which we covered in detail in *Chapter 2, JIT, Hotspot, and GraalJIT*), on Graal JIT (which we covered in *Chapter 4, Graal Just-In-Time Compiler*), or as a native image with Graal AOT (which we covered in *Chapter 5, Graal Ahead-of-Time Compiler and Native Image*). With GraalVM 21.0, Java on Truffle is the new runtime, which can run Java. It is codenamed Espresso. This is still in the *experimental* phase and is not production-ready at the time of writing this book. In this section, we will understand how to run Java applications with this new runtime, and how this can help polyglot programming.

Espresso is a cut-down version of JVM but implements all the core components of JVM, such as the bytecode interpreter, bytecode verifier, Java Native Interface, the Java Debug Wire Protocol, and so on. Espresso reuses all the classes and native libraries from GraalVM. Espresso implements the **JRE (Java Runtime Environment)** library `libjvm.so` APIs. The following figure shows the Espresso stack architecture:

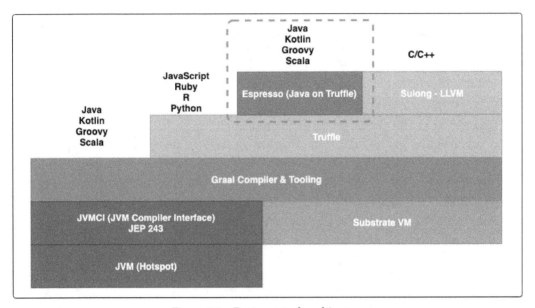

Figure 8.1 – Espresso stack architecture

The figure shows how Espresso is implemented on top of Truffle.

Why do we need Java on Java?

Running Java on Truffle (Espresso) is counter-intuitive, and you might wonder about the advantage of running Java on Truffle, which adds an additional layer on top of Graal. The following are some of the advantages of running Espresso:

- **Hotswap methods, lambdas, and access modifiers at runtime/debug time**: Espresso provides a way to hot-swap methods, lambdas, and access modifiers at runtime during debugging. This is a great feature for developers as it allows them to change the code completely while debugging, and without stopping the runtime and recompiling, the changes take effect at runtime. This speeds up the developer's workflow and increases productivity. It also helps the developers experiment and try things out before committing the code.

- **A sandbox to run untrusted Java code**: Espresso runs like a sandbox on top of Truffle, which can run with access restrictions. This is a great way to run untrusted Java code, by providing specific access. Please refer to the *JavaScript embedded code in Java* section in *Chapter 7, GraalVM Polyglot - JavaScript and Node.js,* to understand more about how to configure the access restrictions.

- **Interoperability between JVM and non-JVM using the same memory space**: Before Espresso, passing data between Java applications and non-JVM dynamic guest languages was not done in the same memory space. This might be because of the performance impact. With Espresso, we can pass data between Java and non-JVM guest languages in the same memory space. This increases the performance of the application.

- **Leveraging Truffle tools and instrumentation**: Running Java on Truffle will help in using all the analysis, diagnosis, and debugging tools developed with Truffle instrumentation. (Refer to the *Understanding Truffle instrumentation* section of *Chapter 6, Truffle for Multi-language (Polyglot) support.*)

- **Ahead-of-time compilation**: Espresso is fully built-in in Java, and so are Truffle and Graal. This enables Espresso to embed into the native image, just like other Truffle guest language interpreters do. This helps in running dynamic code (such as Reflection, JNI, and so on), which was one of the limitations of building native images (refer to the *Native image configuration* section of *Chapter 5, Graal Ahead-of-Time Compiler and Native Image*, for more details). We can now look at segregating the code that requires these dynamic features, and run it on Truffle, while the other parts of the code can be built to `native-image`.

- **Running mixed versions of Java**: Espresso provides the required isolation layer to run Java code that is written in Java 8 to run on Java 11. The Java 8 code can be run on Espresso, which might be running on GraalVM Java 11. This helps in running older code, without changing it, and could be a step in carefully modernizing the code, instead of the big-bang modernization approach that we adopt when we move from the older version of Java to the newer version of Java.

Let's now install and run simple Java code on Espresso.

Installing and running Espresso

Espresso is an optional runtime; it has to be downloaded and installed separately using the Graal Updater tool. Here is the command to install Espresso:

```
gu install espresso
```

To test whether Espresso is installed, let's execute a simple `HelloEspresso.java` application. It is a very simple `Hello World` program, which prints a message. Check out the following code for `HelloEspresso.java`:

```
public class HelloEspresso {
    public static void main(String[] args) {
        System.out.println("Hello Welcome to Espresso!!!");
    }
}
```

Let's compile this application using `javac` and run it with the following command:

```
javac HelloEspresso.java
```

To run Java on Truffle, we just have to pass `-truffle` as a command-line argument to `java`. After running this, we should see the following output:

```
java -truffle HelloEspresso
Hello Welcome to Espresso!!!
```

This validates the installation. We can also use the `-jar` argument along with `-truffle` to run a JAR file. Now let's explore the polyglot capabilities of Espresso.

Exploring polyglot interoperability with Espresso

Espresso is built on Truffle and implements the Truffle polyglot and interoperability APIs. In this section, we will explore these features.

Before we start using the polyglot features, we have to install Espresso polyglot features. To install Espresso polyglot features, we may have to download the Espresso JAR file. You can find the latest version at `https://www.oracle.com/downloads/graalvm-downloads.html`.

The following screenshot shows the JAR file that we will have to download, at the time of writing the book:

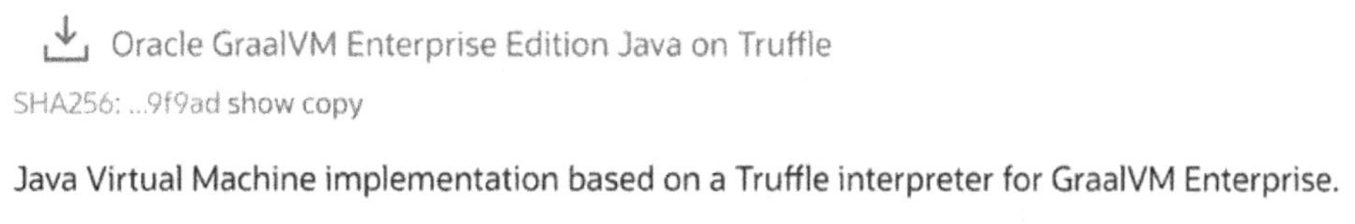

⬇ Oracle GraalVM Enterprise Edition Java on Truffle

SHA256: ...9f9ad show copy

Java Virtual Machine implementation based on a Truffle interpreter for GraalVM Enterprise.

Figure 8.2 – Java on Truffe JAR file download

Once we download this file, we can install it by running the following command:

```
sudo gu install -L espresso-installable-svm-svmee-java11-
darwin-amd64-21.0.0.2.jar
```

Once the installation is successful, we have to rebuild the `libpolyglot` native image, to include Espresso libraries. This library is required to run polyglot support:

```
sudo gu rebuild-images libpolyglot -cp ${GRAALVM_HOME}/lib/
graalvm/lib-espresso.jar
```

This will rebuild the `libpolyglot` native image. We are now ready to use the polyglot capabilities of Espresso. Let's explore these features in the following section.

Exploring Espresso interoperability with other Truffle languages

As you're now aware, Espresso implements the Truffle implementation framework and the `com.oracle.truffle.espresso.polyglot.Polyglot` class implements the polyglot in Espresso. Like any other guest language, we use `-polyglot` in the command-line argument to let Truffle know how to create the polyglot context. Espresso injects a `Polyglot` object into the code, which can be used to interoperate with other languages. Let's explore polyglot programming with Espresso by running the following code:

```
import com.oracle.truffle.espresso.polyglot.Polyglot;
public class EspressoPolyglot {
    public static void main(String[] args) {
        try {
            Object hello = Polyglot.eval("js",
                "print('Hello from JS on Espresso');");
        } catch (Exception e) {
            e.printStackTrace();
        }
    }
}
```

Let's understand the preceding code. The `Polyglot` object provides context for running dynamic languages. The `Polyglot.eval()` method runs foreign language code. The first parameter suggests that it is JavaScript code, and the second parameter is the actual JavaScript code that we want to execute. Let's compile this code with the following command:

```
javac -cp ${GRAALVM_HOME}/languages/java/lib/polyglot.
jar  EspressoPolyglot.java
```

In this command, we are explicitly passing the `polyglot.jar` file in the `-cp` argument (`CLASSPATH`). `polyglot.jar` has all the polyglot implementation of Espresso, including the `com.oracle.truffle.espresso.polyglot.Polyglot` import.

Let's now run the Java application on Espresso. We should pass the `-truffle` argument to run it on Espresso, if we don't do that, it runs on Host JVM. We can see the following output:

```
java -truffle --polyglot EspressoPolyglot
[To redirect Truffle log output to a file use one of the
following options:
* '--log.file=<path>' if the option is passed using a guest
language launcher.
* '-Dpolyglot.log.file=<path>' if the option is passed using
the host Java launcher.
* Configure logging using the polyglot embedding API.]
Hello from JS on Espresso
```

Similarly, we can call other language code. Java is a typed language, unlike other dynamically typed languages on Truffle. When we exchange data between Espresso (Java on Truffle) and other dynamically typed languages such as JavaScript, Python, and so on, we need a way to cast the data types. The `polyglot` object provides a way to cast the data with the `Polyglot.cast()` method. Let's use a simple application to understand how to cast the data, with the following code:

```
import com.oracle.truffle.espresso.polyglot.Polyglot;
import com.oracle.truffle.espresso.polyglot.Interop;
```

Import the `Polyglot` and `Interop` classes. The `Polyglot` class helps us to run guest languages and the `Interop` class implements the Truffle interoperability API, which abstracts the data types between guest languages. Truffle defines an interoperability protocol that provides a clear specification on how the data and message (method calls) exchange happens between Truffle languages, tools, and embedders:

```java
public class EspressoPolyglotCast {
    public static void main(String[] args) {
        try {
            Object stringObject = Polyglot.eval("js",
                "'This is a JavaScript String'");
            Object integerObject = Polyglot.eval("js",
                "1000");
            Object doubleObject = Polyglot.eval("js",
                "10.12345");
            Object arrayObject = Polyglot.eval("js",
                "[1234, 10.2233, 'String element',400,500,
                    'Another Sttring element']");
            Object booleanObject = Polyglot.eval("js",
                "10 > 5");
```

In the preceding code snippet, we are evaluating various JavaScript code snippets that return a string, integer, double, an `array` of integers and a `boolean` value. These values are assigned to a generic `Object`, and then later cast to the respective Java type `String`, `Integer`, `Double`, `Integer[]`, and `Boolean` objects using `Polyglot.cast()`, as observed in the following code snippet:

```java
            String localStringObject =
                Polyglot.cast(String.class, stringObject);
            Integer localIntegerObject =
                Polyglot.cast(Integer.class, integerObject);
            Double localDoubleObject =
                Polyglot.cast(Double.class, doubleObject);
            Boolean localBooleanObject =
                Polyglot.cast(Boolean.class, booleanObject);
            System.out.println("\nString Object : "
                + localStringObject
                    + ", \nInteger : " + localIntegerObject
                    + ", \nDouble : " + localDoubleObject
                    + ", \nBoolean : " + localBooleanObject);
```

Next, we'll print the values. To handle arrays, let's use the `Interop` class to get information about the array object, such as the size of the array with `Interop.getArraySize()`, and iterate through the array with `Interop.readArrayElement()`. Interop also provides a way to check the type of the object and extract the value in a specific data type. In our example, we have evaluated a JavaScript array that has a sequence of integer, double, and string objects. We will use `Interop.fitsInInt()`, `Interop.fitsInDouble()`, and `Interop.isString()` methods to check the types, and accordingly extract the values using `Interop.asInt()`, `Interop.asDouble()`, and `Interop.asString()` methods. The following is the code snippet:

```java
            long sizeOfArray =
                Interop.getArraySize(arrayObject);
            System.out.println(
            "\n Array of objects with Size : " + sizeOfArray );
            for (int i=0; i<sizeOfArray; i++) {
                Object currentElement =
                    Interop.readArrayElement
                    (arrayObject, i);
                if (Interop.fitsInInt(currentElement)) {
                    System.out.println("Integer Element: "
                        +Interop.asInt(currentElement));
                }
                if (Interop.fitsInDouble(currentElement)) {
                    System.out.println("Double Element: "
                        + Interop.asDouble(currentElement));
                }
                if (Interop.isString(currentElement)) {
                    System.out.println("String Element: "
                        + Interop.asString(currentElement));
                }
            }
        } catch (Exception e) {
            e.printStackTrace();
        }
    }
}
```

These values are then printed. Let's compile and run this application. The following is the output:

```
javac -cp ${GRAALVM_HOME}/languages/java/lib/polyglot.
jar EspressoPolyglotCast.java
espresso git:(main) java -truffle --polyglot
EspressoPolyglotCast
String Object : This is a JavaScript String,
Integer : 1000,
Double : 10.12345,
Boolean : true

Array of objects with Size : 6
Integer Element: 1234
Double Element: 1234.0
Double Element: 10.2233
String Element: String element
Integer Element: 400
Double Element: 400.0
Integer Element: 500
Double Element: 500.0
String Element: Another String element
```

In the output, we can see how a dynamically cast language (JavaScript) is captured in a generic `Object` and later cast to specific types. We can also use `Polyglot.isForeignObject(<object>)` to check whether the passed object is a local object or a foreign object.

We saw how we can call other Truffle languages from Espresso, the same way other languages are invoked with `Context polyglot = Context.newBuilder().allowAllAccess(true).build()` and using bindings (refer to the *Bindings* section of *Chapter 7, GraalVM Polyglot - JavaScript and Node.js*) to exchange data and invoke methods.

Java on Truffle Espresso is in very early releases and is at an experimental stage at the time of writing this book. There are a lot of limitations at present, such as a lack of support for the JVM Tool Interface and Java Management Extensions. There are even a lot of performance issues at this point. Please refer to `https://www.graalvm.org/reference-manual/java-on-truffle/` for the latest updates.

Let's now look at two of the most important languages for machine learning – Python and R.

Understanding GraalPython – the Python Truffle interpreter

GraalVM provides a Python runtime. The Python runtime is 3.8 version-compliant and is still in the *experimental* phase at the time of writing this book. In this section, we will install and understand how Python runs on Truffle and Graal. We will also build some sample code, to understand the interoperability features of Graal Python.

Installing Graal Python

Graal Python is an optional runtime and is not installed by default along with GraalVM. To download it, you have to use the Graal Updater tool. The following command downloads and installs Graal Python:

```
gu install python
```

To validate the installation, let's run simple Python code. The following is the source code of `HelloGraalPython.py`:

```
print("Hello Graal Python")
```

It's a very simple Hello World application where we are printing the message. Let's run this application using `graalpython`:

```
graalpython HelloGraalPython.py
```

When we execute the preceding command, we should see the output shown next:

```
graalpython HelloGraalPython.py
Hello Graal Python
```

The preceding output shows that the application is running, and `graalpython` is working.

`graalpython` also supports a virtual environment. The following command will create a virtual environment:

```
graalpython -m venv <name-of-virtual-env>
```

This command will create a virtual environment directory, which will be an isolated environment. GraalPython also comes with `ginstall`, a tool to install supported libraries. The following command will install `numpy` for `graalpython`. `pip` can also be used to install libraries:

```
graalpython -m ginstall install numpy
```

Let's now understand how the GraalPython compilation and interpreter pipeline works.

Understanding the graalpython compilation and interpreter pipeline

Graalpython is a slightly different compilation/interpreter pipeline. To improve the performance of parsing, Graalpython uses an intermediate representation called **Simple Syntax Tree (SST)** and **Scope Tree (ST)**. SST is a simpler representation of the source file, mirroring the source. Typically, when SST is translated to AST, one node in SST may translate to multiple nodes in AST. ST captures the scope information of variables and functions. Together SST and ST are serialized to a `.pyc` file, after the parsing. This is done to speed up the parsing. The next time we run the Python program, `Graalpython` looks for the `.pyc` file and validates whether the file exists, and if it matches the Python source code, then it will deserialize that to build the SST and ST. Otherwise, it will do a full parsing using ANTLR. The following figure shows the full flow. The diagram does not capture all the details. Refer to the *Exploring the Truffle Interpreter/Compiler pipeline* section in *Chapter 6, Truffle for Multi-language (Polyglot) support,* for a more detailed explanation on how Truffle interpreters and Graal JIT execute the code:

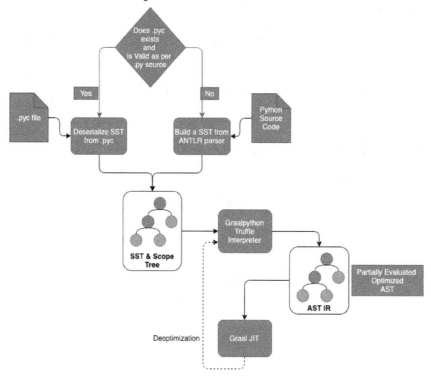

Figure 8.3 – Graalpython compilation/interpreter pipeline

Once the SST and ST are created, they are then converted to an AST intermediate representation and optimized. The final specialized AST is submitted to GraalJIT for further execution after partial evaluation, and the usual flow continues, as explained in *Exploring the Truffle interpreter/compiler pipeline* section of *Chapter 6, Truffle for Multi-language (Polyglot) support*.

So far, we have learned how to run Python programs with GraalPython and how GraalPython optimizes the parsing and optimizes the code using Truffle and GraalJIT. Let's now explore the polyglot interoperability features of GraalPython.

Exploring interoperability between Java and Python

In this section, we will explore the interoperability between Java and Python with sample Java code. The following code calculates the sum of Fibonacci numbers. This class has a findFibonacci() method, which takes in the number of Fibonacci numbers we need and returns an array of those Fibonacci numbers:

```java
public class FibonacciCalculator{
    public int[] findFibonacci(int count) {
        int fib1 = 0;
        int fib2 = 1;
        int currentFib, index;
        int [] fibNumbersArray = new int[count];
        for(index=2; index < count; ++index ) {
            currentFib = fib1 + fib2;
            fib1 = fib2;
            fib2 = currentFib;
            fibNumbersArray[index - 1] = currentFib;
        }
        return fibNumbersArray;
    }
    public static void main(String args[]) {
        FibonacciCalculator fibCal =
            new FibonacciCalculator();
        int[] fibs = fibCal.findFibonacci(10);
    }
}
```

Let's now call the `findFibonacci()` method from Python code. The following is the Python code for calling the method and iterating through the array that is returned by the Java class:

```
import java
import time
fib = java.type("FibonacciCalculator")()
result = fib.findFibonacci(10)
print("Fibonacci number ")
for num in result:
    print(num)
```

In the preceding code, we are using `java.type()` to load the Java class, and we are directly using the returned value as a Python object to call the `findFibonacci()` method, by passing a parameter. We are then able to parse through the result that is returned by the method. Let's compile the Java code and run the Python code. The following shows the terminal output:

```
javac FibonacciCalculator.java
graalpython --jvm --vm.cp=. FibCal.py
Fibonacci number
0
1
2
3
5
8
13
21
34
```

We can see that we are able to call the Java method and get an array of integers and iterate through that, without any extra code for conversion.

Now let's create a simple Python function that uses NumPy to do some quick analysis on a dataset. NumPy is a high-performing Python library for array/matrix manipulations and is widely used in machine learning. To appreciate the value of Graal polyglot, imagine a use case where we have a dataset that has information about various heart attack cases, organized by age, sex, cholesterol levels, chest pain level, and so on, and we want to understand what the average age of the people who had a heart attack after level 3 (high) chest pain is. That is what we will build in this section, to understand the polyglot interoperability between Java and Python, and how we can use the NumPy Python library.

We will use the dataset that is provided on Kaggle on heart attack analysis (`https://www.kaggle.com/rashikrahmanpritom/heart-attack-analysis-prediction-dataset`). This dataset has information about the various heart attack cases, with age, cholesterol levels, sex, chest pain levels, and so on. Here is the Python code to perform the analysis:

```python
import site
import numpy as np
import polyglot as poly
def heartAnalysis():
    heartData = np.genfromtxt('heart.csv', delimiter=',')
    dataOfPeopleWith3ChestPain =
        heartData[np.where(heartData[:,2]>2)]
    averageAgeofPeopleWith3ChestPain =
        np.average(dataOfPeopleWith3ChestPain[:,0])
    # Average age of people who are getting level 3 and
    greater chest pain
    return averageAgeofPeopleWith3ChestPain
poly.export_value("hearAnalysis", heartAnalysis)
```

In the preceding code, we are loading the CSV file into a matrix. Here, we are particularly interested in the third column (indexed as 2). We are loading all the rows where the third column value is greater than 2, and storing it in another variable. We are then averaging that matrix and returning it. This would have taken a lot of code if we had to do the same in Java. Now, let's call this code from Java.

In the following Java code, we will be importing the function definition using the key through the `Binding` object. Here's the complete Java code:

```java
public class NumPyJavaExample {
    public void callPythonMethods() {
        Context ctx =
        Context.newBuilder().allowAllAccess(true).build();
        try {
            File fibCal = new File("./numpy-example.py");
            ctx.eval(Source.newBuilder("python",
                fibCal).build());
            Value hearAnalysisFn =
                ctx.getBindings("python")
                    .getMember("heartAnalysis");

            Value heartAnalysisReport =
                hearAnalysisFn.execute();
            System.out.println(
                "Average age of people who are getting level 3
                    and greater chest pain :" +
                        heartAnalysisReport.toString());

        } catch (Exception e) {
            System.out.println("Exception : " );
            e.printStackTrace();
        }
    }
    public static void main(String[] args) {
        NumPyJavaExample obj = new NumPyJavaExample();
        obj.callPythonMethods();
    }
}
```

In the previous Java code, we are creating a `Context` object and evaluating the Python code in `numpy-example.py`. We are then accessing the function definition through binding and invoking the Python function and are able to get the value. We are printing the value that is returned. The following is the output of running this Java code:

```
$ java NumPyJavaExample
Average age of people who are getting level 3 and greater chest
pain :55.869565217391305
```

In the preceding output, we can see that the first call took time, however, the subsequent calls took almost no time at all to execute. This not only demonstrates how we can interoperate with Python code from Java code but also how Truffle and Graal optimize the execution.

In this section, we explored Java and Python interoperability. In the next section, we will explore interoperability between dynamic languages with Python.

Exploring interoperability between Python and other dynamic languages

To explore interoperability between Python and other dynamic languages, let's use the same `numpy-example.py` that we used in the previous section. Let's call this method from JavaScript.

The following is the JavaScipt that calls the Python code:

```
function callNumPyExmple() {
    Polyglot.evalFile('python', './numpy-example.py');
    heartAnalysis = Polyglot.import('heartAnalysis');
    result = heartAnalysis();
    return result;
}
result = callNumPyExmple();
print ('Average age of people who are getting level 3 and
    greater chest pain : '+  String(result));
```

In the previous code, we can see how we are importing the Python `heartAnalysis()` function in JavaScript using the `Polyglot.import()` function. This returns the average value that we are printing. Let's run this code, and we can see the following result:

```
$ js --polyglot numpy-caller.js
Average age of people who are getting level 3 and greater chest
pain : 55.869565217391305
```

Let's now create JavaScript code, which will have functions to calculate squares. To demonstrate how JavaScript code can be called from Python, here's the JavaScript code:

```
var helloMathMessage = " Hello Math.js";
function square(a) {
    return a*a;
}
Polyglot.export('square', square);
Polyglot.export('message', helloMathMessage)
```

It's a very simple JavaScript function that returns the square of the passed value. We are also exporting the `square()` function and a variable message, which carries the value of the `helloMathMessage` variable.

Now let's invoke this method from Python code. The following is the Python code that will import and invoke the preceding JavaScript methods:

```
import polyglot
polyglot.eval(path="./math.js", language="js")
message = polyglot.import_value('message')
square = polyglot.import_value('square')
print ("Square numbers by calling JS->Python: " +
    str(square(10, 20)))
print ("Hello message from JS: " + message)
```

In this code, we are using the Python `polyglot` object to evaluate the JavaScript file. We then imported all the exported functions/variables by calling the `polyglot.import_value()` function, by using the same key used by JavaScript to export functions or variables. We are then able to invoke those functions and access the `message` variable and print the values. The following output is what you get after you run the preceding code:

```
$ graalpython --jvm --polyglot mathUser.py
Square numbers by calling JS->Python: 100
Hello messagr from JS:  Hello Math.js
```

We can see how Python code is importing and invoking JavaScript code. This demonstrates two-way interoperability. The code is very similar to other languages, such as R and Ruby.

In this section, we explored and gained a good understanding of how the Python interpreter works with Truffle to run optimally on GraalVM. Let's now explore and understand the R language interpreter on GraalVM.

Understanding FastR – the R Truffle interpreter

GraalVM provides an R Truffle interpreter for a GNU-compatible R runtime. This runtime supports R programs and **REPL** (**read-eval-print-loop**) mode, where we can rapidly test the code while we write the code interactively. FastR is the project that developed this R runtime.

Installing and running R

Just like Graal Python, the R runtime does not come with GraalVM by default. We have to download and install it using Graal Updater. Use the following command to download and install R and Rscript:

```
gu install r
```

To run R, we need the OpenMP runtime library. This can be installed using `apt-get install libcomp1` on Ubuntu and `yum install libcomp` on Oracle Linux. The library is installed in macOS by default. Apart from this, you will need C/C++/Fortran, if the R code has C/C++/Fortran code. R is also in the experimental phase at the time of writing this book, so not everything is supported yet. Please refer to the GraalVM documentation (`https://docs.oracle.com/en/graalvm/enterprise/20/docs/reference-manual/r/`) for the latest information.

Let's now test R. To explore the R interpreter, let's run it in interactive mode. The following terminal output shows the interactive mode to test R installation:

```
R
R version 3.6.1 (FastR)
Copyright (c) 2013-19, Oracle and/or its affiliates
Copyright (c) 1995-2018, The R Core Team
Copyright (c) 2018 The R Foundation for Statistical Computing
Copyright (c) 2012-4 Purdue University
Copyright (c) 1997-2002, Makoto Matsumoto and Takuji Nishimura
All rights reserved.


FastR is free software and comes with ABSOLUTELY NO WARRANTY.
You are welcome to redistribute it under certain conditions.
Type 'license()' or 'licence()' for distribution details.


R is a collaborative project with many contributors.
Type 'contributors()' for more information.


Type 'q()' to quit R.
[Previously saved workspace restored]
```

We see that we are using the FastR GraalVM version from the version numbers listed in the preceding output. Let's now test whether our FastR interpreter is working by running some Python commands as shown next:

```
> 1+1
[1] 2
> abs(-200)
[1] 200
```

We can see that it is interactively providing the results. Let's now just plot a simple example. The best way is to call `example()`, which will show the plot, as shown next:

```
> example (plot)

plot> require(stats) # for lowess, rpois, rnorm

plot> plot(cars)

plot> lines(lowess(cars))
NULL

plot> plot(sin, -pi, 2*pi) # see ?plot.function
NULL

plot> ## Discrete Distribution Plot:
plot> plot(table(rpois(100, 5)), type = "h", col = "red", lwd =
10,
plot+        main = "rpois(100, lambda = 5)")
NULL

plot> ## Simple quantiles/ECDF, see ecdf() {library(stats)} for
a better one:
plot> plot(x <- sort(rnorm(47)), type = "s", main = "plot(x,
type = \"s\")")

plot> points(x, cex = .5, col = "dark red")
```

This will result in a pop-up window with the plotted graph. The following figure shows a screenshot of the graph that popped up:

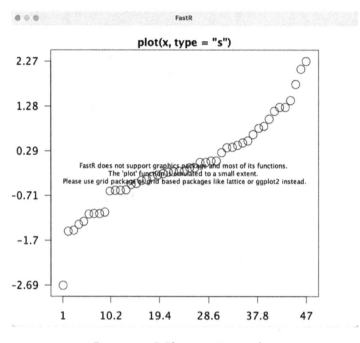

Figure 8.4 – R Plot output screenshot

At the time of writing this book, some warnings appeared while running the preceding plot commands. These warnings list some of the limitations of FastR. However, this might change in upcoming versions. The following are the warnings that popped up:

```
NULL
Warning messages:
1: In lines.default(lowess(cars)) :
  lines.default not supported. Note: FastR does not support
graphics package and most of its functions. Please use grid
package or grid based packages like lattice instead.
2: In plot.function(sin, -pi, 2 * pi) :
  plot.function not supported. Note: FastR does not support
graphics package and most of its functions. Please use grid
package or grid based packages like lattice instead.
3: In axis(...) :
  axis not supported. Note: FastR does not support graphics
package and most of its functions. Please use grid package or
grid based packages like lattice instead.
```

```
4: In points.default(x, cex = 0.5, col = "dark red") :
   points.default not supported. Note: FastR does not support
graphics package and most of its functions. Please use grid
package or grid based packages like lattice instead.
>
```

Now that we can see R is working fine, let's now explore the interoperability features of FastR.

Exploring the interoperability of R

In this section, to explore polyglot and interoperability with R, we will run some inline JavaScript and also load sample JavaScript code and import the exported functions and variables. We will use R interactive mode to do this so that it's easy to understand. To run R in polyglot mode, we have to pass the --polyglot argument. The following is the command:

```
R --polyglot
```

This will start the R runtime in interactive mode with the following output:

```
R version 3.6.1 (FastR)
Copyright (c) 2013-19, Oracle and/or its affiliates
Copyright (c) 1995-2018, The R Core Team
Copyright (c) 2018 The R Foundation for Statistical Computing
Copyright (c) 2012-4 Purdue University
Copyright (c) 1997-2002, Makoto Matsumoto and Takuji Nishimura
All rights reserved.


FastR is free software and comes with ABSOLUTELY NO WARRANTY.
You are welcome to redistribute it under certain conditions.
Type 'license()' or 'licence()' for distribution details.


R is a collaborative project with many contributors.
Type 'contributors()' for more information.


Type 'q()' to quit R.
[Previously saved workspace restored]


>
```

Now, let's start with simple inline JavaScript:

```
> x <- eval.polyglot('js','[100,200,300,400]')
> print(x)
[polyglot value]
[1] 100 200 300 400
> print(x[3])
[1] 300
```

In the preceding interactive session, we are calling the eval.polyglot() function, with the language ID and the expression. In this case, we are specifying it as JavaScript with a language ID of js and then passing an array of elements. Then we are printing the array and the third element in the array. The eval.polyglot() function provides the polyglot context and runs other language code. Now let's load a simple JavaScript code file. The following is the code for math.js:

```
var helloMathMessage = " Hello Math.js";
function add(a, b) {
    print("message from js: add() called");
    return a+b;
}
function subtract(a, b) {
    print("message from js: subtract() called");
    return a-b;
}
function multiply(a, b) {
    print("message from js: multiply() called");
    return a*b;
}
Polyglot.export('add', add);
Polyglot.export('subtract', subtract);
Polyglot.export('multiply', multiply);
Polyglot.export('message', helloMathMessage)
```

The preceding code is very straightforward. We have defined the add(), subtract(), and multiply() functions and a simple variable, message, which has a string value, Hello Math.js. We are then using Polyglot.export() to export it for other languages to have access to these functions and the variable.

Now let's load this JavaScript file and execute the exported code; we will be running the instructions in interactive mode. You'll find the interactive session here, with an explanation of what we are doing:

```
> mathjs <- eval.polyglot('js', path='/chapter8/r/math.js')
```

This instruction loads the JavaScript file. Make sure that the path is updated with the exact path where you have the JavaScript file. Now let's import the exported functions and variable into R:

```
> message <- import('message')
> add <- import('add')
> subtract <- import('subtract')
> multiply <- import('multiply')
```

In the preceding instructions, we are using the import () function to import the exported functions and variables. It is very important to use the same string that we used to export in the JavaScript file. These imports are assigned to a variable. Now let's call these functions and print the variable:

```
> add(10,20)
message from js: add() called
[1] 30
> subtract(30,20)
message from js: subtract() called
[1] 10
> multiply(10,40)
message from js: multiply() called
[1] 400
> print(message)
[1] " Hello Math.js"
>
```

As you can see, we can call the JavaScript functions and print the variable. This demonstrates how we can use JavaScript but we can similarly use all other Truffle languages. Let's now explore how to access a Java class from R. Here is the code for the `HelloRPolyglot` class, which we will be calling from R:

```java
import org.graalvm.polyglot.Context;
import org.graalvm.polyglot.Value;


public class HelloRPolyglot {
    public String hello(String name) {
        System.out.println("Hello Welcome from hello");
        return "Hello Welcome from hello " + name;
    }


    public static void helloStatic() {
        System.out.println("Hello from Static hello()");
        try {
            Context polyglot = Context.create();
            Value array = polyglot.eval("js",
                "print('Hello from JS inline in HelloRPolyglot
                    class')");
        } catch (Exception e) {
            e.printStackTrace();
        }
    }
    public static void main(String[] args) {
        HelloRPolyglot.helloStatic();
    }
}
```

Let's understand the preceding code. We have a static method, `helloStatic()`, that calls inline JavaScript, which prints a message, and we have another method, `hello()`, that takes an argument and prints a `hello` message.

```
javac HelloRPolyglot.java
java HelloRPolyglot
Hello Welcome to R Polyglot!!!
Hello from JS inline in HelloRPolyglot class
```

Now that the class is working fine, let's start the R interactive mode. This time, we have to pass the --jvm argument to let the R runtime know that we will be using Java, and also pass the --vm argument, to set CLASSPATH to the current directory where we have the Java class file:

```
R --jvm --vm.cp=.
R version 3.6.1 (FastR)
Copyright (c) 2013-19, Oracle and/or its affiliates
Copyright (c) 1995-2018, The R Core Team
Copyright (c) 2018 The R Foundation for Statistical Computing
Copyright (c) 2012-4 Purdue University
Copyright (c) 1997-2002, Makoto Matsumoto and Takuji Nishimura
All rights reserved.

FastR is free software and comes with ABSOLUTELY NO WARRANTY.
You are welcome to redistribute it under certain conditions.
Type 'license()' or 'licence()' for distribution details.

R is a collaborative project with many contributors.
Type 'contributors()' for more information.

Type 'q()' to quit R.
[Previously saved workspace restored]

>
```

Now that the R is loaded, let's run the instructions to call the `hello()` method in the Java class. We use the `java.type()` function to load the class. The following is the interactive session:

```
> class <- java.type('HelloRPolyglot')
> print(class)
[polyglot value]
$main
[polyglot value]

$helloStatic
[polyglot value]

$class
[polyglot value]
```

In the preceding interactive session, we can see that the class is loaded successfully, and when we print the class, we see that it lists the various methods in it. Now let's create an instance of this class. We use the `new()` function to do that. The following is the output of the interactive session with the `new()` function:

```
> object <- new(class)
> print(object)
[polyglot value]
$main
[polyglot value]

$helloStatic
[polyglot value]

$class
[polyglot value]

$hello
[polyglot value]
```

In the preceding code, we can see that the object is successfully created, as it prints all the methods in the class. Now let's call these methods. We will use the class to call the static method and object to call `hello()`, by passing a parameter. The following is the output of the interactive session:

```
> class$helloStatic()
Hello from Static heloo()
Hello from JS inline in HelloRPolyglot class
NULL
> object$hello('FastR')
Hello Welcome from hello
[1] "Hello Welcome from hello FastR"
>
```

In the preceding session, we can see the output of calling both the methods.

Let's take a real-life example of how we can use the power of plotting a graph using R and use the plotted graph in Node.js. Earlier in the chapter, we used a dataset that we got from Kaggle that has heart attack data. Let's use that dataset to plot a graph comparing the ages of people and their cholesterol levels on a web page that is generated by Node.js.

Let's initialize a Node.js project with npm init. The following is the output console where we are providing the name of the project and other project parameters:

```
$ npm init
This utility will walk you through creating a package.json
file.
It only covers the most common items, and tries to guess
sensible defaults.
See `npm help init` for definitive documentation on these
fields and exactly what they do.
Use `npm install <pkg>` afterwards to install a package and
save it as a dependency in the package.json file.
Press ^C at any time to quit.
package name: (plotwithr-node)
version: (1.0.0)
description:
entry point: (plotWithR.js)
test command:
git repository:
```

```
keywords:
author:
license: (ISC)
About to write to /Users/vijaykumarab/AB-Home/Developer/
GraalVM-book/Code/chapter8/r/plotWithR-node/package.json:
{
  "name": "plotwithr-node",
  "version": "1.0.0",
  "description": "",
  "main": "plotWithR.js",
  "scripts": {
    "test": "echo \"Error: no test specified\" && exit 1"
  },
  "author": "",
  "license": "ISC"
}
Is this OK? (yes)
```

This should generate a Node.js boilerplate. We will need the Express.js library to expose a REST endpoint. Let's now install the express library and use `--save` to update the `package.json` file with the dependency. Here's the output:

```
$ npm install express --save
added 50 packages, and audited 51 packages in 2s
found 0 vulnerabilities
```

Let's now write the Node.js code to load the dataset (`heart.csv`) and render a bar chart as a `scalar vector graph` (SVG). To plot, we will be using the Lattice package (you can find more details about this library at `https://www.statmethods.net/advgraphs/trellis.html`).

So, here's the Node.js code:

```
const express = require('express')
const app = express()
app.get('/plot', function (req, res) {
  var text = ""
  text += Polyglot.eval('R',
    `svg();
    require(lattice);
    data <- read.csv("heart.csv", header = TRUE)
    print(barchart(data$age~data$chol,
        main="Age vs Cholestral levels"))
    grDevices:::svg.off()
    `);
  res.send(text)
})
app.listen(3000, function () {
  console.log('Plot with R -  listening on port 3000!')
})
```

Let's go through the code to understand it. We are loading Express.js and defining a '/plot' endpoint. We are using Polyglot.eval() to run our R code. We are initializing the SVG and loading the Lattice package. We are then loading the heart.csv file and rendering the graph as a bar chart, and then adding the SVG response, generated to the HTML as a response for the /plot endpoint.

Let's now run this code. The following shows the output after running the code:

```
node --jvm --polyglot plotWithR.js
Plot with R - listening on port 3000!
Loading required package: lattice
```

Go to `http://locahost:3000/plot` to invoke the endpoint, on a browser. The following figure shows a screenshot of the output:

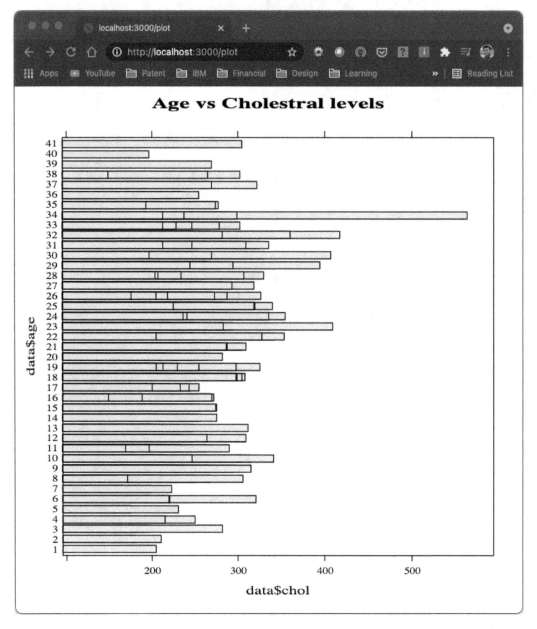

Figure 8.5 – Output of calling /plot

R is a very powerful language for statistical computations and machine learning. This opens up opportunities for us to embed R code or call R code within the same runtime, from various other languages. If we had to do the same logic in Java, it might take a lot of effort.

Summary

In this chapter, we went into the details of how Python, R, and Java on Truffle interpreters are implemented in Truffle. We also explored the polyglot interoperability features that these languages provide, along with coding examples. We understood the differences in the way each of these languages is interpreted. The chapter provided a hands-on walkthrough of how to run code and write polyglot applications in these various languages. We used very simple code so that you could easily understand the concepts and API to implement polyglot applications.

You should be able to use this knowledge to write polyglot applications on GraalVM. Though most of these languages are still in the experimental phase at the time of writing the book, they provide great opportunities to build high-performance polyglot applications.

In the next chapter, you will gain good hands-on experience and understanding of how polyglot works, how to build Python and R applications on GraalVM, and how to interoperate between these programs. You will also gain a good knowledge of GraalVM's new runtime, Java on Truffle.

Questions

1. What is Java on Truffle?

2. What are the advantages of Java on Truffle?

3. What is the use of the `Polyglot.cast()` method?

4. What are SST and ST?

5. What is a `.pyc` file?

6. What is the polyglot binding method used to exchange data and function definitions in GraalPython?

7. How can you import other language definitions in R?

8. How can you load a Java class in R?

Further reading

- GraalVM Enterprise Edition: `https://docs.oracle.com/en/graalvm/enterprise/19/index.html`

- GraalVM Language Reference: `https://www.graalvm.org/reference-manual/languages/`.

9
GraalVM Polyglot – LLVM, Ruby, and WASM

In the previous chapter, we covered Truffle interpreters for Java, Python, and R and interoperability between languages. In this chapter, we will cover other languages' implementations, such as the following:

- **LLVM**: The LLVM Truffle interpreter
- **TruffleRuby**: The Ruby language interpreter implementation
- **WebAssembly (WASM)**: WebAssembly implementation

All of these language implementations are still in the *experimental* phase and are not released for production, at the time of writing this book. However, we will explore the features and build some code to understand the various concepts.

In this chapter, we will cover the following topics:

- Understanding LLVM, Ruby, and WASM interpreters and their polyglot features
- Understanding the compatibility and limitations of these various language interpreters

By the end of this chapter, you will have had hands-on experience in building polyglot applications with LLVM, Ruby, and WASM interpreters.

Technical requirements

This chapter requires the following to follow along with the various coding/hands-on sections:

- The latest version of GraalVM.

- Various language Graal runtimes. We will cover in the chapter how to install and run these runtimes.

- Access to GitHub. There are some sample code snippets that are available on the Git repository. The code can be downloaded from the following link: `https://github.com/PacktPublishing/Supercharge-Your-Applications-with-GraalVM/tree/main/Chapter09`.

- The Code in Action video for this chapter can be found at `https://bit.ly/3hT7Z1A`.

Understanding LLVM – the (Sulong) Truffle interface

LLVM is a compiler infrastructure that provides a modular, reusable set of compiler components that can form a toolchain to compile source code to machine code. The toolchain provides various levels of optimization, on an **intermediate representation** (**IR**). Any source language can use this toolchain, as long as the source code can be represented as an LLVM IR. Once the source code is represented as an LLVM IR, that language can utilize the advanced optimization techniques that LLVM provides. You can refer to the LLVM project at `https://llvm.org/`. There are various compilers that are already built on this infrastructure. Some of the most popular ones are Clang (for C, C++, and Objective C), Swift (used extensively by Apple), Rust, and Fortran.

Sulong is an LLVM interpreter that is written in Java and internally uses the Truffle language implementation framework. This enables all language compilers that can generate LLVM IR to directly run on GraalVM. The following diagram shows how Sulong enables LLVM languages to run on GraalVM:

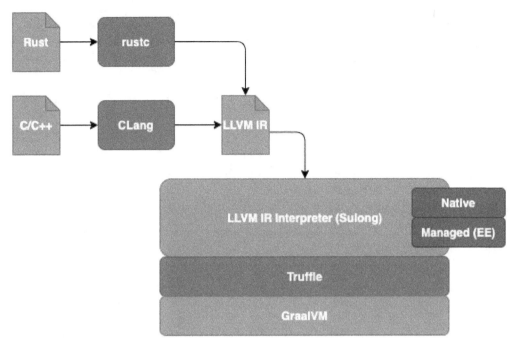

Figure 9.1 – LLVM compilation pipeline

The preceding figure shows how LLVM works at a very high level. C/C++ source code is compiled to LLVM IR. Let's understand this diagram better:

- C/C++ code is compiled by Clang to LLVM IR.

- The lli GraalVM LLVM IR interpreter has two versions: *native* and *managed*. Native is the default lli version that comes both in Community Edition and Enterprise Edition. Managed lli is only available in Enterprise Edition and provides a managed mode of execution, which we will be covering in the *Understanding the LLVM managed environment* section.

- The LLVM IR interpreter performs the initial optimization and integrates with Truffle and Graal for further optimizations at runtime.

To understand how LLVM IR looks, here is some example C code where we are adding two numbers and returning the result:

```
int addInt(int a, int b)
{
    return a + b;
}
```

When we pass this through Clang, the following LLVM IR is generated. We can generate the LLVM IR using the `clang -S -emit-llvm cfile.c` command:

```
define dso_local i32 @addInt(i32 %0, i32 %1) #0 !dbg !7 {
  %3 = alloca i32, align 4
  %4 = alloca i32, align 4
  store i32 %0, i32* %3, align 4
  call void @llvm.dbg.declare(metadata i32* %3,
  metadata !12, metadata!DIExpression()), !dbg !13
  store i32 %1, i32* %4, align 4
  call void @llvm.dbg.declare(metadata i32* %4,
  metadata !14, metadata !DIExpression()), !dbg !15
  %5 = load i32, i32* %3, align 4, !dbg !16
  %6 = load i32, i32* %4, align 4, !dbg !17
  %7 = add nsw i32 %5, %6, !dbg !18
  ret i32 %7, !dbg !19
}
```

The preceding IR clearly shows how the code gets converted into **static single assignment** (**SSA**) from the a variable (`%0`), which is passed as a parameter, and b (`%1`) is allocated `%3` and `%4` respectively. The values are then loaded into `%5` and `%6` respectively and added to `%7`. The value in `%7` is returned. SSA makes the optimization algorithms easy to implement, as the algorithms don't have to keep track of the value changes in a single variable, since every change in the variable is assigned to a single static value.

The LLVM IR can be passed to Sulong, which is the LLVM IR interpreter, which internally uses the Truffle language implementation framework to run the code on GraalVM. This takes advantage of the advanced optimization features of GraalVM.

Installing the LLVM toolchain

LLVM is available as an optional runtime, and can be installed using Graal Updater, with the following command:

```
gu install llvm-toolchain
```

LLVM gets installed in `$GRAALVM_HOME/languages/llvm/native/bin`. We can check the path by using the following command:

```
$GRAALVM_HOME/bin/lli --print-toolchain-path
```

This will print the path where LLVM is installed. Enterprise Edition also comes with a managed LLVM. We will cover that later in this chapter.

The GraalVM LLVM runtime can execute language code that is converted to LLVM bitcode. The GraalVM `lli` tool interprets bit code and then compiles dynamically using Graal **just in time** (**JIT**). `lli` also enables interoperability with dynamic languages. The syntax for the `lli` command is shown next:

```
lli [LLI options] [GraalVM options] [polyglot options] <bitcode
file> [program args]
```

`lli` can execute plain bitcode or native executables with embedded bitcode (Linux: ELF and macOS: Mach-O).

Let's quickly verify the installation with some simple code:

```
#include <stdio.h>

int main() {
    printf("Welcome to LLVM Graal \n");
    return 0;
}
```

Let's compile the code with Clang:

```
clang HelloGraal.c -o hellograal
```

This runs the `hellograal` application. Let's run it using `lli`. `lli` is the LLVM interpreter:

```
lli hellograal
```

The following shows the output:

```
lli hellograal
Hello from GraalVM!
```

We can directly execute /hellograal and get the same output. This is called native execution. Sometimes native executions run faster than lli, but we don't get the polyglot features that GraalVM provides with lli. Let's take something more complex; let's convert FibonacciCalculator.java to C. Here is the source code in C:

```c
#include <stdio.h>
#include <stdlib.h>
#include <sys/time.h>
long fib(int i) {
    int fib1 = 0;
    int fib2 = 1;
    int currentFib, index;
    long total = 0;
    for (index = 2; index < i; ++index)
    {
        currentFib = fib1 + fib2;
        fib1 = fib2;
        fib2 = currentFib;
        total += currentFib;
    }
    printf("%ld \n", total);
    return total;
}
int main(int argc, char const *argv[])
{
    for (int i = 1000000000; i < 1000000010; i++)
    {
        struct timeval tv_start;
        struct timeval tv_end;
        long time;
        gettimeofday(andtv_start, NULL);
        fib(i);
        gettimeofday(andtv_end, NULL);
        time = (tv_end.tv_sec*1000000 +
            tv_end.tv_usec) - (tv_start.tv_sec*1000000 +
                tv_start.tv_usec);
        printf("i=%d time: %10ld\n", i, time);
```

```
    }
    return 0;
}
```

Let's create an executable by running the following command:

```
/Library/Java/JavaVirtualMachines/graalvm-ee-java11-21.0.0.2/
Contents/Home/languages/llvm/native/bin/clang
FibonacciCalculator.c -o fibonacci
```

We have to make sure we use the right version of Clang. We have to use the Clang that is provided with the GraalVM LLVM toolchain; otherwise, we won't be able to use `lli`. If we use normal Clang and try to execute the generated binary with `lli`, we get the following error:

```
oplevel executable /fibonacci does not contain bitcode
        at <llvm> null(Unknown)
```

Once the binary file is created, we execute it with `lli` to use the GraalVM JIT compilation capabilities. Here is the output when executed with `lli`:

```
  llvm git:(main) lli fibonacci
-24641037439717
i=1000000000 time:    5616852
-24639504571562
i=1000000001 time:    5592305
-24640314634125
i=1000000002 time:    5598246
-24639591828533
i=1000000003 time:    1116430
-24639679085504
i=1000000004 time:    1092585
-24639043536883
i=1000000005 time:    1140553
-24638495245233
i=1000000006 time:    1117817
-24637311404962
i=1000000007 time:    1121831
-24635579273041
```

```
i=1000000008 time:        1103494
-24636958268145
i=1000000009 time:        1109705
```

Let's plot these results on a graph and see how the performance improved over iterations. The following is the graph:

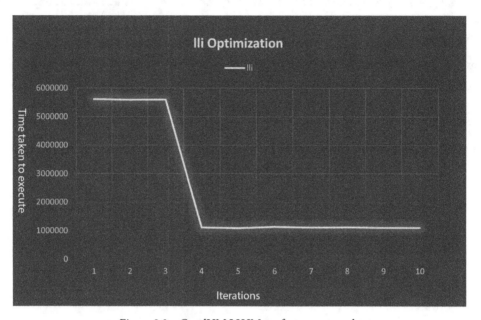

Figure 9.2 – GraalVM LLVM performance graph

We can see a significant improvement over iterations. We can see that initially, it runs slow, but after the third iteration, it improves by almost sixfold.

Exploring LLVM interoperability

In this section, we will explore the interoperability features of LLVM and see how we can interact between Java, LLVM (C), and JavaScript.

Java and LLVM interoperability

Let's first try to call the FibonacciCalculator.c executable from Java. Here is the source code for FibonacciCalculatorLLVMEmbed.java:

```java
import java.io.File;
import org.graalvm.polyglot.Context;
import org.graalvm.polyglot.Source;
```

```
import org.graalvm.polyglot.Value;


public class FibonacciCalculatorLLVMEmbed {
    public static void main(String[] args) {
        try {
            Context polyglot = Context.newBuilder()
                .allowAllAccess(true).build();
            File file = new File("fibpoly");
            Source source = Source.newBuilder("llvm",
                file).build();
            Value fibpoly = polyglot.eval(source);
            fibpoly.execute();
        } catch (Exception e) {
            e.printStackTrace();
        }
    }
}
```

The source code is very similar to what we did with JavaScript. We are creating a `Context` object, loading the compiled file with a `Source` object, and building the `Source` object from a binary using `SourceBuilder`. The `Context` object is then used to evaluate the `Source` object and is finally executed.

Let's compile this Java file and run it with GraalVM Java. The following code shows the output of the file:

```
java FibonacciCalculatorLLVMEmbed
Inside C code: 10944
Returned value to Java 10944
```

We can see that we are able to call C code from Java. In this case, we just executed the C application. Let's now try to call the member function directly and pass an integer, and get the total as a `long` type.

Here is the modified C code:

```
#include <stdio.h>
#include <stdlib.h>
#include <sys/time.h>
long fib(int i) {
```

```
    int fib1 = 0;
    int fib2 = 1;
    int currentFib, index;
    long total = 0;
    for (index = 2; index < i; ++index) {
        currentFib = fib1 + fib2;
        fib1 = fib2;
        fib2 = currentFib;
        total += currentFib;
    }
    printf("Inside C code: %ld \n", total);
    return total;
}
```

The preceding code implements the FibonacciCalculator logic that we had previously done using Java.

In this section, we looked at how to invoke a C method from Java. Let's now understand how to invoke C code from a dynamic language such as JavaScript.

Exploring JavaScript and LLVM interoperability

Let's check the interoperability with JavaScript. JavaScript provides a very simple snippet. Here is the JavaScript code (FibonacciCaller.js):

```
var fibpoly = Polyglot.evalFile("llvm" , "fibpoly");
var fib = fibpoly.fib(20);
print("Returned value to JS: "+ fib);
```

Let's run this JavaScript as follows:

```
js --polyglot FibonacciCaller.js
Inside C code: 10944
Returned value to JS: 10944
```

We can see that we are now able to pass data to C code and execute the C method.

Understanding the LLVM managed environment

GraalVM Enterprise Edition provides a managed environment of LLVM. We can find a version of the LLVM managed toolchain under the /languages/llvm/managed directory, while the default, unmanaged toolchain can be found under /languages/llvm/native.

When the managed version of lli is used, the managed environment or managed mode of execution can be enabled with the --llvm.managed flag. In this section, let's understand what the managed mode of execution is, and why it is specifically required for LLVM.

To understand the problems we typically face, let's take some very simple code, ManagedLLVM.c. In the following code, we are intentionally trying to copy a character array to an uninitialized char pointer:

```c
#include <stdlib.h>
#include <stdio.h>
int main() {
  char *string;
  char valueStr[10] = "Hello Graaaaaal";
  strcpy(string, valueStr);
  printf("%s", string);
  free(string);
  return 0;
}
```

Let's compile this code; Clang actually warns us about incorrectly initializing valueStr. valueStr is only defined to take 10 characters, but we are assigning more than 10 characters. Let's assume that we ignore the warnings and proceed. The application is still built and can be executed. The following is the output of compiling the ManagedLLVM.c file:

```
/Library/Java/JavaVirtualMachines/graalvm-ee-java11-21.0.0.2/
Contents/Home/languages/llvm/native/bin/clang ManagedLLVM.c -o
managedllvm
ManagedLLVM.c:5:23: warning: initializer-string for char array
is too long
  char valueStr[10] = "Hello Graaaaaal";
                      ^~~~~~~~~~~~~~~~~~
```

```
ManagedLLVM.c:6:3: warning: implicitly declaring library
function 'strcpy' with type
      'char *(char *, const char *)' [-Wimplicit-function-
      declaration]
  strcpy(string, valueStr);
  ^
```

```
ManagedLLVM.c:6:3: note: include the header <string.h> or
explicitly provide a declaration for 'strcpy'
```

```
2 warnings generated.
```

```
ManagedLLVM.c:5:23: warning: initializer-string for char array
is too long
  char valueStr[10] = "Hello Graaaaaal";
                      ^~~~~~~~~~~~~~~~~
```

```
ManagedLLVM.c:6:3: warning: implicitly declaring library
function 'strcpy' with type
      'char *(char *, const char *)' [-Wimplicit-function-
      declaration]
  strcpy(string, valueStr);
  ^
```

```
ManagedLLVM.c:6:3: note: include the header <string.h> or
explicitly provide a declaration for 'strcpy'
```

```
2 warnings generated.
```

If we ignore the warnings and still run the application binary, we obviously get a page fault. This kills the host process and stops the application completely. Such problems cause a core dump and crash the application, and there are a lot of such instances with languages such as C/C++ where we face these kinds of issues. The following is the output when we run the code in native mode (directly native and native lli):

```
./managedllvm
```

```
[1]    30556 segmentation fault  ./managedllvm
```

GraalVM LLVM managed execution mode provides a graceful way of handling these issues. Let's take the same code and compile it this time with the managed version of Clang and run it with the managed version of lli. Let's run the application binary with the managed version of lli:

```
llvm git:(main) lli --llvm.managed managedllvm
Illegal null pointer access in 'store i64'.
        at <llvm> main(ManagedLLVM.c:6:112)
```

It still fails, but this time it is not a segmentation fault or crash; it is throwing an exception. Exceptions can be caught and handled gracefully.

To understand how to handle this better, lets create a Java class (`ManagedLLVM.java`) that calls the `managedllvm` executable from Java and handles the exception gracefully:

```java
import java.io.File;
import org.graalvm.polyglot.Context;
import org.graalvm.polyglot.Source;
import org.graalvm.polyglot.Value;
public class ManagedLLVM {
    public static void main(String[] args) {
        try {
            Context polyglot = Context.newBuilder()
                .allowAllAccess(true)
                .option("llvm.managed", "true")
                .build();
            File file = new File("managedLLVM");
            Source source =
                Source.newBuilder("llvm", file).build();
            Value mllvm = polyglot.eval(source);
            mllvm.execute();
        } catch (Exception e) {
            System.out.println("Exception occured....");
            e.printStackTrace();
        }
    }
}
```

Note that we are now creating a `Context` object with the `llvm.manage` option as `true`. That is very critical for us to run the executable in managed execution mode. Let's compile and run this Java application:

```
javac ManagedLLVM.java
java ManagedLLVM
Exception occured....
Illegal null pointer access in 'store i64'.
        at <llvm> main(ManagedLLVM.c:6:112)
```

```
    at org.graalvm.sdk/org.graalvm.polyglot.Value.
    execute(Value.java:455)
    at ManagedLLVM.main(ManagedLLVM.java:13)
```

We can see that the Java application is now able to catch the exception and we could be writing exception handling code here. Moreover, it is not stopping the application. This is one of the greatest features of running LLVM in managed execution mode, and it is supported even in polyglot environments.

Ruby is another language that has a high-performance interpreter implementation in GraalVM. Let's explore and understand TruffleRuby, the Ruby Truffle implementation, in the next section.

Understanding TruffleRuby – the Ruby Truffle interpreter

TruffleRuby is a high-performance implementation of the Ruby programming language on GraalVM that is built on Truffle. In this section, we will explore some of the language-specific concepts, with code examples, to gain a good understanding of Ruby implementation on GraalVM.

Installing TruffleRuby

TruffleRuby, too, does not come by default with GraalVM installation. You'll have to download and install it using the Graal updater tool. To install TruffleRuby, use the following command:

```
gu install ruby
```

After installing Ruby, we have to run some post-install scripts to make OpenSSL C extensions work. We need to run post_install_hook.sh, which you will find under the ruby/lib/truffle directory. Let's test the installation with a simple Ruby application:

```
print "enter a "
a = gets.to_i
print "Enter b "
b = gets.to_i
c = a + b
puts "Result " + c.to_s
```

The preceding code accepts the values of a and b as integers from the user, adds the numbers, and prints the result as a string. This is a very simple Ruby application to test Ruby. Let's run this program on TruffleRuby. The following is the terminal output:

```
truffleruby helloruby.rb
enter a 10
Enter b 20
Result 30
```

Now that we know that TruffleRuby is installed and working, let's understand how the TruffleRuby interpreter works.

Understanding the TruffleRuby interpreter/compiler pipeline

TrufflyRuby, like any other guest language, is a Truffle interpreter implementation. The following figure shows the TruffleRuby interpreter/compiler pipeline:

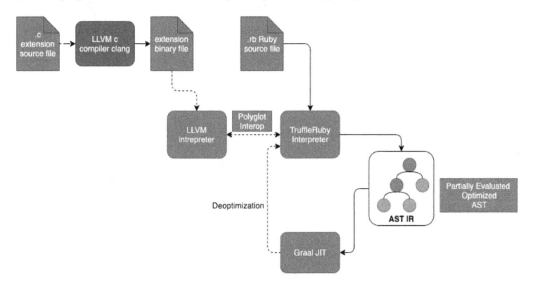

Figure 9.3 – TruffleRuby compiler/interpreter pipeline

The interpreter/compiler pipeline is very similar to other guest languages. The preceding diagram does not capture all the details. Refer to the *Exploring the Truffle interpreter/compiler pipeline* section in *Chapter 6, Truffle for Multi-language (Polyglot) support,* for a more detailed explanation on how Truffle interpreters and Graal JIT executes the code. The TruffleRuby interpreter builds the AST after parsing and performs optimizations and submits the code to Graal JIT, like any other guest language. However, one of the key differences is the way it handles C extensions. C extensions are an integral part of Ruby programming, and traditionally, C extensions are plugged into the Ruby interpreter. TruffleRuby handles this using the LLVM interpreter. This naturally provides polyglot interoperability and we can use other LLVM languages such as C++, Rust, and Swift, not just C. TruffleRuby brings in polyglot interoperability as it's built on Truffle. Let's explore the polyglot interoperability features of TruffleRuby.

Exploring Polyglot interoperability with TruffleRuby

TruffleRuby brings all the polyglot interoperability features of Truffle and implements similar APIs. Let's explore these features in this section. Let's write some simple Ruby code that calls the JavaScript file we built in previous sections that exports simple math functions and variables:

```
var helloMathMessage = " Hello Math.js";
function add(a, b) {
    print("message from js: add() called");
    return a+b;
}
function subtract(a, b) {
    print("message from js: subtract() called");
    return a-b;
}
function multiply(a, b) {
    print("message from js: multiply() called");
    return a*b;
}
Polyglot.export('add', add);
Polyglot.export('subtract', subtract);
Polyglot.export('multiply', multiply);
Polyglot.export('message', helloMathMessage);
```

The following is the Ruby code that demonstrates the polyglot capabilities:

```
arrayPy = Polyglot.eval("python", "[10, 10.23456,
    'Array String element']")
puts arrayPy.to_s
lengthOfArray = arrayPy.size
puts "Iterating through the Python Array object of size "
    + lengthOfArray.to_s
for i in 0..lengthOfArray - 1
    puts "Element at " + i.to_s + " is " + arrayPy[i].to_s
end
```

In the preceding code, we are calling JavaScript code that creates an array of three elements: an integer, a float, and a string. We are evaluating this JavaScript code in line with `Polyglot.eval()`. We are then iterating through the array and printing the values. In the following code, we are loading `math.js` and importing the `message` variable and the `add()`, `subtract()`, and `multiply()` functions. We are then invoking those functions and printing the results:

```
Polyglot.eval_file("./math.js")
message = Polyglot.import("message")
addFunction = Polyglot.import_method("add")
subtractFunction = Polyglot.import_method("subtract")
multiplyFunction = Polyglot.import_method("multiply")


puts "Message from JS " + message
puts "Result of add(10,20) " + add(10,20).to_s
puts "Result of subtract(10,20) " + subtract(40,20).to_s
puts "Result of multiply(10,20) " + multiply(10,20).to_s
```

Now let's run this code. The following is the output:

```
truffleruby --polyglot mathJsCaller.rb
#<Python [10, 10.23456, 'Array String element']>
Iterating throught the Python Array object of size 3
Element at 0 is 10
Element at 1 is 10.23456
Element at 2 is Array String element
Message from JS  Hello Math.js
```

```
message from js: add() called
Result of add(10,20) 30
message from js: subtract() called
Result of subtract(10,20) 20
message from js: multiply() called
Result of multiply(10,20) 200
```

We can see that we are able to iterate through the JavaScript array, and also call the functions in the math.js JavaScript code.

Let's now explore how to interoperate with Java. We use the Java.type() method to load the Java class. Let's use a slightly modified version of FibonacciCalculator. Here's the Java source code:

```java
public class FibonacciCalculator{
    public int[] findFibonacci(int count) {
        int fib1 = 0;
        int fib2 = 1;
        int currentFib, index;
        int [] fibNumbersArray = new int[count];
        for(index=2; index < count+1; ++index ) {
            currentFib = fib1 + fib2;
            fib1 = fib2;
            fib2 = currentFib;
            fibNumbersArray[index - 1] = currentFib;
        }
        return fibNumbersArray;
    }
    public void iterateFibonacci() {
        long startTime = System.currentTimeMillis();
        long now = 0;
        long last = startTime;
        for (int i = 1000000000; i < 1000000010; i++) {
            int[] fibs = findFibonacci(i);
            long total = 0;
            for (int j=0; j<fibs.length; j++) {
                total += fibs[j];
            }
```

```
            now = System.currentTimeMillis();
            System.out.printf("%d (%d ms)%n", i , now - last);
            last = now;
        }
        long endTime = System.currentTimeMillis();
        System.out.printf("total: (%d ms)%n",
            System.currentTimeMillis() - startTime);


    }
    public static void main(String args[]) {
        FibonacciCalculator fibCal =
            new FibonacciCalculator();
        fibCal.iterateFibonacci();
    }
}
```

We have defined two methods: `findFibonacci()` and `iterateFibonacci()`. The `findFibonacci()` method takes an integer and returns the Fibonacci numbers, as per the requested count. `iterateFibonacci()` iterates and generates a large number of Fibonacci numbers, and times it to check how the code performs.

The following code is the Ruby script to load the `FibonacciCalculator` class and call the `findFibonacci(int count)` method. We pass an integer value to this method and we get an array of integers. We then go through the array and print the values. We are also calling `iterateFibonacci()` to compare how it performs with running directly with Java:

```
fibclass = Java.type('FibonacciCalculator')
fibObject = fibclass.new
fibonacciArray = fibObject.findFibonacci(10)
for i in 0..fibonacciArray.size - 1
    puts "Element at " + i.to_s + " is " + fibonacciArray[i]
    .to_s
end
puts "Calling iterateFibonacci()"
fibObject.iterateFibonacci()
```

To run this Ruby script, we need to pass --jvm in the command line along with --vm.cp to point to the path where the Java class is available. The following is the output of running TruffleRuby:

```
truffleruby --jvm --vm.cp=. fibonacciJavaCaller.rb
Element at 0 is 0
Element at 1 is 1
Element at 2 is 2
Element at 3 is 3
Element at 4 is 5
Element at 5 is 8
Element at 6 is 13
Element at 7 is 21
Element at 8 is 34
Element at 9 is 55
Calling iterateFibonacci()
1000000000 (2946 ms)
1000000001 (1011 ms)
1000000002 (1293 ms)
1000000003 (1016 ms)
1000000004 (1083 ms)
1000000005 (1142 ms)
1000000006 (1072 ms)
1000000007 (994 ms)
1000000008 (982 ms)
1000000009 (999 ms)
total: (12538 ms)
```

We can see how we are able to call the Java class and go through the Java array in Ruby, and even `iterateFibonacci()` performed reasonably well. Let's try to compare this with running this Java class directly with Java. Here's the output:

```
java FibonacciCalculator
1000000000  (2790 ms)
1000000001  (592 ms)
1000000002  (1120 ms)
1000000003  (927 ms)
1000000004  (955 ms)
1000000005  (952 ms)
1000000006  (974 ms)
1000000007  (929 ms)
1000000008  (923 ms)
1000000009  (924 ms)
total:  (11086 ms)
```

We can see that the performance of Ruby is on par with running Java directly. TruffleRuby is one of the best-performing Ruby runtimes. TruffleRuby is in the experimental phase, while writing this book; for the latest information, refer to `https://www.graalvm.org/reference-manual/ruby/`.

One of the biggest advantages of using Ruby is RubyGems. Ruby has a vast library, which the developer community has built over a period of time. All Gems are hosted in https://rubygems.org/. With GraalVM Polyglot, this opens up a huge opportunity to use these gems in Java or any other language, supported by GraalVM. To illustrate this, let's use a gem in a Java program. There is a gem called math_engine (`https://rubygems.org/gems/math_engine`). This has a very interesting method to evaluate complex mathematical expressions. Let's assume that we are building a complex algebra calculator that can be used to evaluate complex expressions. Let's use this gem in a Ruby program, and invoke it from Java.

Let's first install the gem. To install the gem, let's use Bundler (https://bundler.io/). Bundler is a package manager (equivalent to npm in Node.js). To install Bundler, use the gem install command. The following is the output of installing Bundler:

```
gem install bundler
Fetching bundler-2.2.17.gem
Successfully installed bundler-2.2.17
1 gem installed
```

Let's now create a Gemfile. Bundler uses the configuration in a Gemfile to install all the packages/gems. (This is equivalent to package.json in npm.) Here is the source of the Gemfile:

```
source 'https://rubygems.org'
gem 'math_engine'
```

We are providing the source of the Gem repository and specifying the gems that depend on our Ruby module. Let's now run Bundler to install these gems. (The bundle install command should be executed in the folder where we have the Ruby program and Gemfile.)

The bundle install command will install all the gems. Let's now use the math_engine gem, and define a method called eval() in Ruby that takes in the expression, evaluates it, and returns the result:

```
require 'rubygems'
require 'math_engine'
def eval(exp)
    engine = MathEngine.new
    ret = engine.evaluate(exp)
    puts(ret)
    return ret.truncate(4).to_f()
end
Polyglot.export_method('eval')
```

In the preceding source code, we are exporting the method using Polyglot.export_method(), so that it can be accessed by other languages. Let's now call this eval() method from a Java program.

The following is the Java source code:

```java
public class MathEngineExample {
    public void evaluateExpression(String exp) {
        Context ctx = Context.newBuilder()
            .allowAllAccess(true).build();
        try {
            File fibCal =
                new File("./math_engine_expression.rb");
            ctx.eval(Source.newBuilder("ruby",
                fibCal).build());
            Value evaluateFunction =
                ctx.getBindings("ruby").getMember("eval");
            Double evaluatedValue =
                evaluateFunction.execute(exp).asDouble();
            System.out.printf("Evaluated Expression : "
                + evaluatedValue.toString());
        } catch (Exception e) {
            System.out.println("Exception : " );
            e.printStackTrace();
        }
    }
    public static void main(String[] args) {
        MathEngineExample obj = new MathEngineExample();
        obj.evaluateExpression("20 * (3/2) + (5 * 5)
            / (100.5 * 3)");
    }
}
```

In the previous Java code, we are using the Context object to load the Ruby runtime and our Ruby program. We are then binding the eval() method and executing it with the passed expression. The value is then captured and converted to a string for printing. In the main() method, we are passing a complex mathematical expression. Let's now compile and run this Java code. The following is the output:

```
java MathEngineExample
0.30082918739635157545605306799e2
Evaluated Expression : 30.0829
```

The first output comes from Ruby put_s() and the next output is coming from Java. This opens up a huge opportunity to use the vast library of gems.

Understanding TruffleRuby Vs CRuby Vs JRuby

Ruby has a lot of implementations on the market. JRuby and CRuby are two popular implementations of Ruby.

JRuby is the Ruby programming language implemented in Java. CRuby is the Ruby programming language implemented in C. The following diagram shows the high-level compilation pipeline of JRuby and CRuby. JRuby is one of the highest-performing implementations of Ruby, as it brings in the optimizations and JIT compilation. JRuby also does not have a global interpreter lock, like in CRuby, and this allows concurrent execution, and hence is faster. However, JRuby starts slow but performs better over a period of time:

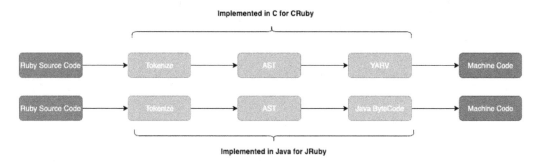

Figure 9.4 – JRuby and CRuby compilation pipeline

TruffleRuby outperforms JRuby and CRuby. Please refer to https://www.graalvm. org/ruby/ for more detailed optcarrot and Rubycon benchmark results.

Understanding GraalWasm – the WASM Truffle interpreter

GraalVM provides an interpreter and compiler for WASM code called GraalWasm. GraalWasm opens up possibilities of building polyglot web applications that perform close to natively. Before we get into the details of GraalWasm, let's have a quick overview of WASM.

Understanding WASM

WASM is a binary format that can run on most modern browsers at near-native speeds. Web applications have become more and more sophisticated and demand a high-performance, near-native experience. JavaScript can only get to a certain level, and we have seen a lot of very good applications built on JavaScript that provide almost native experience. WASM augments JavaScript and other technologies to allow us to compile C, C++, and Rust programs on the web. The following figure shows a very simple pipeline of building WASM applications:

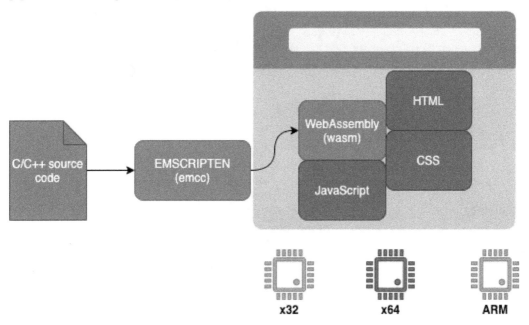

Figure 9.5 – WASM compilation pipeline flow

In the preceding figure, we can see how C/C++ code can be compiled to WASM using Emscripten, and it coexists with the other web technologies and runs on web browsers. Both JavaScript and WASM execute logic on the browser. The main difference is WASM is delivered as a binary that is already optimized at compile time. There is no need for abstract syntax trees and type specializations and speculations (refer to the *Exploring the Truffle interpreter/compiler pipeline* section in *Chapter 6, Truffle for Multi-language (Polyglot) support*, to understand more about AST and speculative optimization).

This is one of the reasons why WASM has a smaller footprint and faster performance. Modern web application architectures leverage WASM to perform more advanced computational logic in the browser, while JavaScript is used to run the user interface and simple application logic. GraalWasm opens up even more possibilities by bringing in polyglot interoperability and embedding. Let's explore that in the next section.

Understanding GraalWasm architecture

Let's understand how GraalWasm works. The following figure shows the compilation pipeline for GraalWasm. The diagram does not capture all the details. Refer to the *Exploring the Truffle interpreter/compiler pipeline* section in *Chapter 6, Truffle for Multi-language (Polyglot) support*, for a more detailed explanation on how Truffle interpreters and Graal JIT execute code:

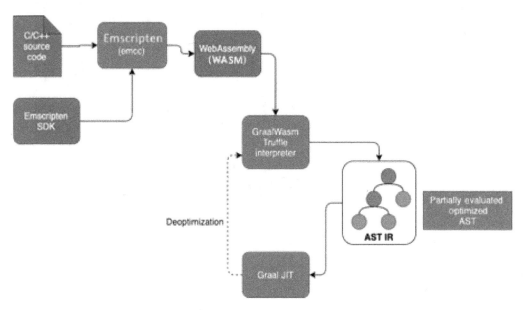

Figure 9.6 – GraalWasm compilation/interpreter pipeline

C/C++ source code is compiled using Emscripten (emcc). Emscripten is a drop-in replacement for gcc or Clang, built on LLVM. Emscripten compiles the source to a WASM (.wasm) file. The GraalWasm interpreter creates the AST and performs optimizations. There is not much optimization that is required, as the WASM is generated from a strongly typed language, and so already a lot of optimization is performed. Since the WASM format is a sequence of instructions, to reduce the memory footprint, GraalWasm builds an AST, where each node is pointing to a block in WASM (this is called the WASM block node), instead of creating a node for each instruction. The GraalWasm interpreter then optimizes the AST and passes it to Graal for execution.

Installing and running GraalWasm

GraalWasm is an optional component; it does not come with the GraalVM installation. We have to install it using Graal Updater. To install GraalWasm, we can use the following command to download and install it:

```
gu install wasm
```

Let's now build a WASM binary and run it with GraalWasm. To compile C code to WASM, we have to install emcc. The following section walks through the steps to install emcc.

Installing Emscripten (emcc)

Emscripten is installed using Emscripten SDK. We can download the SDK from the Git repository. Let's execute the following command on a terminal:

```
git clone https://github.com/emscripten-core/emsdk.git
```

This downloads emsdk, which also has all the required installation scripts. git clone will create an emsdk directory. Let's move that to the cd emsdk folder and perform git pull to make sure it is up to date. We can install the emcc toolchain by executing the following command:

```
./emsdk install latest
```

This will download all the required toolchains and SDKs. Once it's downloaded, we need to activate emsdk and then set the environment, using the following commands in sequence:

```
./emsdk activate latest
./emsdk_env.sh
```

Once all of these commands are successfully executed, we should be able to check whether Emscripten is installed by running the emcc command on the terminal.

Now let's build a WASM of the following C code. This is a slight modification of the code that we have already used in the *Exploring LLVM interoperability* section:

```c
#include <stdio.h>
#include <stdlib.h>
#include <sys/time.h>
static long fib(int i) {
```

```
    int fib1 = 0;
    int fib2 = 1;
    int currentFib, index;
    long total = 0;
    for (index = 2; index < i; ++index) {
        currentFib = fib1 + fib2;
        fib1 = fib2;
        fib2 = currentFib;
        total += currentFib;
    }
    return total;
}
```

We have defined a method that calculates Fibonacci numbers, sums them up, and returns the sum of the Fibonacci numbers. In the following code snippet of the main() method, we are looping through and calling this fib() method in interactions, just to test the method, and printing the total and time taken for executing each iteration:

```
int main(int argc, char const *argv[])
{
    for (int i = 10000; i < 10010; i++) {
        struct timeval tv_start;
        struct timeval tv_end;
        long time;
        gettimeofday(&tv_start, NULL);
        fib(i);
        gettimeofday(&tv_end, NULL);
        time = (tv_end.tv_sec*1000000 + tv_end.tv_usec) -
            (tv_start.tv_sec*1000000 + tv_start.tv_usec);
        printf("i=%d time: %10ld\n", i, time);
    }
    return 0;
}
```

Let's compile this code with emcc with the following command:

```
emcc -o fibonacci.wasm fibonacci.c
```

In the previous command, the -o option is used to specify the name of the output file; in this case, fibonacci.wasm is the binary file that is generated, after successful compilation. Let's run this file with GraalWasm by executing the following command:

```
wasm --Builtins=wasi_snapshot_preview1 fibonacci.wasm
```

In the preceding command, we used --Builtins to pass the wasi_snapshot_preview1 module that the Emscripten toolchain requires. The following shows the output after executing:

```
wasm --Builtins=wasi_snapshot_preview1 fibonacci.wasm
i=10000 time:          6563
i=10001 time:          6519
i=10002 time:          6841
i=10003 time:          8455
i=10004 time:          6838
i=10005 time:          7156
i=10006 time:          7214
i=10007 time:           237
i=10008 time:           265
i=10009 time:           247
```

We can see the output generated instantly and is quite fast for the number of Fibonacci numbers we wanted to create and add up. Also, we can see a huge performance improvement from 6,519 ms to 247 ms.

Summary

In this chapter, we went into the details of how LLVM, Ruby, and WASM, Java on Truffle and Ruby interpreters are implemented on Truffle. We also explored the polyglot interoperability features that these languages provide, along with coding examples. We understood the differences in the way each of these languages is interpreted. The chapter provided a hands-on walkthrough of how to run code and write polyglot applications in these various languages.

You should be able to use this knowledge to write polyglot applications on GraalVM. Though most of these languages are still in the experimental phase at the time of writing the book, they provide great opportunities to build high-performance polyglot applications. In the next chapter, we will see how the new frameworks such as Quarkus and Micronaut implement Graal for most optimum microservices architecture.

Questions

1. What is Sulong?

2. What is the LLVM managed mode of execution?

3. How does TruffleRuby implement C extensions?

4. What is WASM?

5. What is `emcc`?

Further reading

- GraalVM Enterprise Edition: `https://docs.oracle.com/en/graalvm/enterprise/19/index.html`

- GraalVM language reference: `https://www.graalvm.org/reference-manual/languages/`

- GraalWasm announcement blog post: `https://medium.com/graalvm/announcing-graalwasm-a-webassembly-engine-in-graalvm-25cd0400a7f2`

- *WASM engine in GraalVM: Introducing Oracle's GraalWasm*: `https://jaxenter.com/graalvm-webassembly-graalwasm-165111.html`

Section 4: Microservices with Graal

This section will walk you through how Graal can be used to build cloud-native applications as microservices and how this can help in serverless. It will also cover a case study of how Graal can be used along with Quarkus and Micronaut to build a microservices application:

- *Chapter 10, Microservices Architecture with GraalVM*

10
Microservices Architecture with GraalVM

In previous chapters, we looked at how GraalVM builds on top of Java VM and provides a high-performance polyglot runtime. In this chapter, we will explore how GraalVM can be the core runtime for running microservices. A lot of microservices frameworks already run on GraalVM. We will explore some of the popular frameworks and build a sample application with them. We will also explore a serverless framework. We will take a case study and look at how we can architect the solution.

By the end of this chapter, you will have acquired a good understanding of how to package applications as containers, running GraalVM, and how to build microservices applications using Micronaut, Quarkus, and Spring Boot. This chapter expects you to have a fair understanding of the Java programming language and some exposure to building Java microservices.

In this chapter, we will cover the following topics:

- An overview of GraalVM microservices architecture
- An understanding of how GraalVM helps to build microservices architecture
- Building microservices applications
- A case study to help understand how to go about solutioning a microservices application built on GraalVM
- Implementing a microservice with Spring Boot, Micronaut, Quarkus, and the Fn Project serverless framework

Technical requirements

This chapter provides a hands-on guide for building Java microservices. This requires some of the software to be installed and set up. The following is a list of prerequisites:

- **Source code**: All the source code referred to in this chapter can be downloaded from the Git repository at `https://github.com/PacktPublishing/Supercharge-Your-Applications-with-GraalVM/tree/main/Chapter10`.
- **GraalVM**: GraalVM needs to be installed. For detailed instructions on installing and setting up GraalVM, refer to `https://www.graalvm.org/docs/getting-started/#install-graalvm`.
- **Spring Boot**: Refer to `https://spring.io/guides/gs/spring-boot/` for more details on how to set up and use Spring Boot.
- **Micronaut**: We will be building code using the Micronaut framework. Please refer to `https://micronaut.io/download/` for more details on how to download and set up Micronaut.
- **Quarkus**: We will be building microservices using the Quarkus framework. Please refer to `https://quarkus.io/` for more details on how to set up and use Quarkus.
- **fn project**: We will be building a serverless application/function using fn project. Please refer to `https://fnproject.io/` for more details on how to download, install, and set up fn project.
- The Code in Action video for this chapter can be found at `https://bit.ly/3f7iT1T`.

So, let's begin!

Overview of microservices architecture

Microservices are one of the most popular architectural patterns and have been proven to be the best architectural pattern for cloud-native application development. Microservices patterns help to decompose and structure applications into smaller, manageable, and self-contained components that expose functionality through a standard service interface. The following are some of the advantages of microservices architectural patterns:

- **Loose coupling**: Since the application is decomposed into services that provide a standard interface, the application component can be independently managed, upgraded, and fixed without affecting the other dependent components. This helps in easily changing the application logic based on growing business needs and changes.

- **Manageability**: Since the components are self-contained, it is very easy to manage these applications. The components can be owned by smaller squads for development and can be deployed independently without deploying the whole application. This assists with rapid development and deployments using DevOps.

- **Scalable**: Scalability is one of the key requirements of cloud-native applications. Scalability in monoliths is an issue, as we have to scale the whole application, even though we just need to scale some part of the functionality. For example, during high demands, we might want to scale the ordering, shopping cart, and catalog services more than any other functionality of a retail portal. That is not possible in monoliths, but if these components are decomposed into independent microservices, it's easy to scale them individually and set autoscale parameters so that they scale based on demand. This helps in utilizing cloud resources more effectively, at a lower cost.

Let's now explore how GraalVM helps to build high-performance microservices architectures.

Building microservices architecture with GraalVM

GraalVM is ideal for microservices architecture as it helps to build high-performance Java applications with a smaller footprint. One of the most important requirements for microservices architecture is a smaller footprint and faster startup. GraalVM is an ideal runtime for running polyglot workloads in the cloud. There are some cloud-native frameworks already available on the market that can build applications to run optimally on GraalVM, including Quarkus, Micronaut, Helidon, and Spring.

Understanding GraalVM containers

Traditionally, applications are deployed on infrastructure that was pre-configured and set up for the applications to run. The infrastructure consisted of both hardware and a software platform that runs the applications. For example, if we have to run a web application, we will have to set up the operating system (such as Linux or Windows, for example) first. The web application server (Tomcat, WebSphere) and database (such as MySQL, Oracle, or DB2) are set up on a pre-defined hardware infrastructure, and then the applications are deployed on top of these web application servers. This takes a lot of time, and we may have to repeat this approach every time we have to set up the applications.

To reduce the setup time and to make the configurations much easier to manage, we moved to virtualizing the infrastructure by pre-packaging the application, along with various platform components (application servers, databases, and suchlike) and the operating system, into self-contained **Virtual Machines** (**VMs**). (These VMs are not to be confused with **Java Virtual Machine** (**JVM**). JVM is more of a platform for running Java applications. VMs in this context are much more than just an application platform.)

Virtualization helped to solve a lot of configuration and deployment issues. It also allowed us to optimize the usage of hardware resources by running multiple VMs on the same machine and utilizing resources better. VMs are bulky as they come with their own operating system and are tough to rapidly deploy, update, and manage.

Containerization solved this issue by bringing in another layer of virtualization. Most modern architectures are built on containers. Containers are units of software that package code and all the dependencies and environment configurations. Containers are lightweight, standalone executable packages that be deployed on container runtimes. The following diagram shows the difference between VMs and containers:

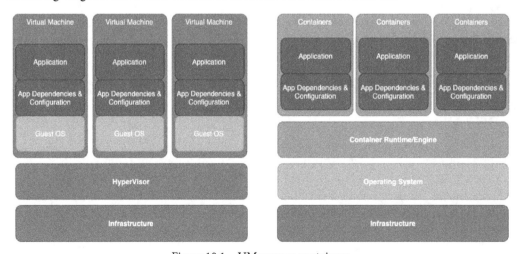

Figure 10.1 – VMs versus containers

GraalVM is a perfect application platform (especially when it is compiled as native code) to be packaged along with the application in the same container. GraalVM provides the smallest footprint and faster startups and execution to rapidly deploy and scale up the application components.

The preceding diagram shows how the application can be containerized with GraalVM. In the previous model, each of the containers has its own VM, which has logic for memory management, profiling, optimization (**JIT**), and so on. What GraalVM provides is a common runtime along with the container runtime, and just the application logic is containerized. Since GraalVM also supports multiple languages and interoperability between these languages, the containers can be running application code that is written in multiple languages.

The following diagram shows the various scenarios of how containers can be deployed with GraalVM:

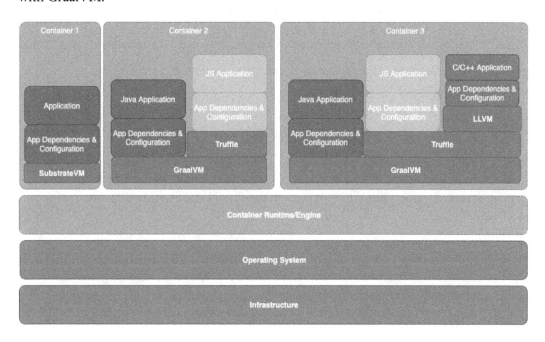

Figure 10.2 – GraalVM container patterns

In the preceding diagram, we can see various configurations/scenarios. Let's go through the details:

- **Container 1**: In this container, we can see a native image running. This is by far the most optimal configuration with the smallest footprint and a faster load.

- **Container 2**: In this container, we have a Java application and a JavaScript application running on Truffle with interoperability.

- **Container 3**: Similar to container 2, we also can see a C/C++ application.

Container 1 is the most optimal configuration for running cloud-native unless we have application code written in a different programming language that needs to interoperate. Another approach is to compile native images and split them into separate containers and use standard protocols such as REST to interact.

These containers can be deployed in the cloud using various orchestrators, such as Docker Swarm, Kubernetes (including Azure Kubernetes Service, AWS Elastic Kubernetes Service, and Google Kubernetes Engine), AWS Fargate, and Red Hat OpenShift.

Let's explore how GraalVM can be used as a common runtime in microservices architecture with the help of a case study.

Case study – online book library

To understand how to implement microservices on GraalVM using various modern microservices frameworks, let's go through a very simple case study. Later in the chapter, we will pick one of the services from this architecture and build it using different frameworks.

This case study involves building a simple website that shows a catalog of books. The catalog lists all the books. You can search and browse the books by specific keywords and should be able to select and obtain more details relating to the book. The user can then select and save it as a wishlist in a library of books. In the future, this can be extended to place an order for this book. But to keep it simple, let's assume that we're searching, browsing, and creating personal libraries in **MVP** (**Minimum Viable Product**) scope. Let's also have a section in the catalog where the user can see a book prediction based on what is in their library. This will help us to do a polyglot with some machine learning code, too.

Functional architecture

Let's go through the thought process of building this application. We will first start by decomposing the functionality. For this, we will require the following services:

- **Catalogue UI Service**: This web page is the home page where the user lands after successfully logging in (we will not be implementing the login, authentication, and authorization in MVP). This web page presents a way to search and view the books. This will be implemented as a micro-frontend (refer to `https://micro-frontends.org/` for more details on micro-frontends). We will have three UI components as follows:

  i. **Book List UI component**: This component shows a list of all the books.

  ii. **Book Details UI component**: This component shows all the details pertaining to the selected book.

  iii. **Predicted Books UI component**: This component shows the books that are predicted, based on the books in the library.

- **Library UI Service**: This lists the books in your personal library and allows the user to add or delete books from this library.

Now, to support these UI services, we will require microservices that store, fetch, and search the books. The following are the services that we will need:

- **Catalogue Service**: These services provide the RESTful APIs to browse, search, and view the book details.

- **Prediction Service**: To demonstrate the polyglot feature of GraalVM, let's assume that we already have machine learning code that we have developed using Python, and that can predict the book, based on the books that are available in the library. We will embed this Python code in this Java microservice to demonstrate how GraalVM can help us to build optimized embedding polyglot applications.

- **Library Service**: This service will provide all the restful APIs for accessing books in the library, as well as for adding and deleting them from the library.

- **Book Info Service**: Let's decide to use the Google Books API (`https://developers.google.com/books`) to get all the details about the books. We will need a service that proxies the Google Books API. This will help us to manage the data that is coming from the Google Books API. This also provides a proxy layer, so that we can always switch to a different Book API service without changing the whole application.

Now we will need storage to store the information about the books that have been added to the personal libraries and to cache the data about the books, for faster fetching (instead of calling the Google Books API every time). To do so, we will require the following data services:

- **User Profile Data**: This stores the user profiles.

- **User Library Data**: This stores the books that the particular user has selected for their library.

- **Book Cache Data**: We will need to cache the book information so that we don't have to call the Google Books API for information that we have already fetched. This will not only improve performance; it will also reduce costs as the Google Books API may charge you for the number of calls made.

The following diagram illustrates how these components work together:

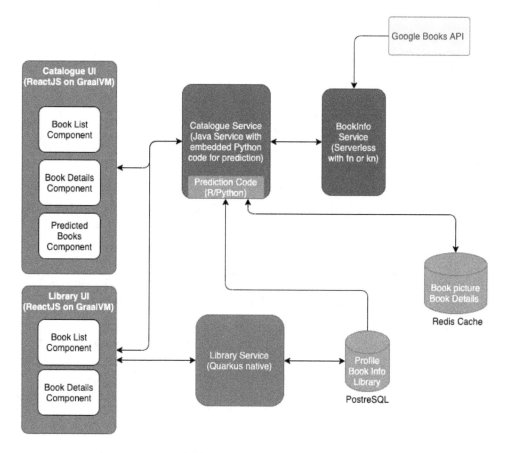

Figure 10.3 – Book library application – functional architecture

We have taken various architectural decisions while building the final architecture. Let's quickly review them:

- **Micro-frontends**: We decided to make the UI components micro-frontends so that it's easier for us to manage and reuse the UI code. As we can see, both the catalog UI and library UI reuse the same components to render the list of books and show the book details. We are choosing ReactJS as this provides a very sound framework for micro-frontend implementation.

- **Embedding Python**: We decided to reuse the Python code already built for prediction. We decided to embed that as part of our catalog service to provide an endpoint that will provide a list of predicted books. This will also help us to demonstrate the capabilities of polyglot. We will use the pure Java implementation of microservices, as most modern microservices frameworks do not support polyglot.

- **Serverless**: We decided to render the book info service *serverless* as it does not need to keep the state; it just calls the Google Books API and passes the information.

- **Book information cache**: We decided to use Redis to store the book information cache so that we don't have to go back to the Google Books API each time, thereby improving performance and reducing the cost of calling Google APIs.

Let's now look at what the deployment architecture will look like on Kubernetes. Please refer to `https://kubernetes.io/` for more details on how Kubernetes orchestrates the containers and provides a scalable and highly available solution. The following section assumes that you have a good understanding of Kubernetes.

Deployment architecture

The containers are deployed on the Kubernetes cluster. The following diagram shows the deployment architecture of these containers in Kubernetes. This can be similar in any cloud:

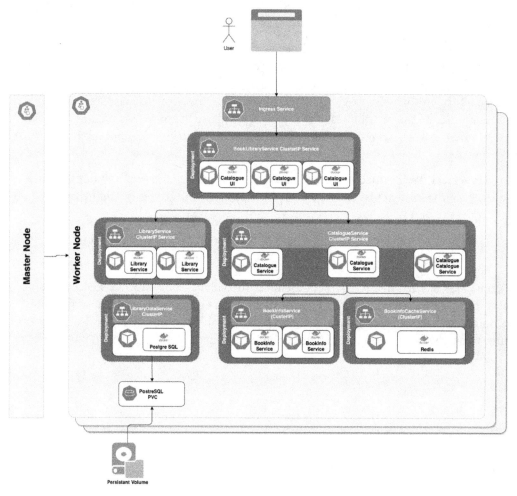

Figure 10.4 – Kubernetes deployment architecture for the book library application

Let's understand the terms used in the preceding diagram in more detail:

- **Ingress Service deployment**: Ingress Service will be the entry to the cluster. This will be a Kubernetes deployment that has the inbound port as 8080, and the target port pointing to the cluster IP of the Catalogue UI page, which is the home page.

- **ClusterUIService deployment**: This deployment has the `reactjs` implementation of the home page and the library page, which internally use the same set of `reactjs` components. This calls Library Service to get the information regarding the books stored in the personal library. This service also calls Catalog Service, which has all the REST endpoints to search and browse the book's details.

- **LibraryService deployment**: Library Service is implemented in Quarkus as a native image and provides the endpoints for accessing personal library information. This uses Library Data Service.

- **LibraryDataService deployment**: Library Data Service is a PostgreSQL container that stores all the user profile and personal library information. It also uses a persistent volume so that when a node goes down, the information is stored.

- **CatalogueInfoService deployment**: This deployment has the implementation of `CatalogueInfoService` in Quarkus native mode. This service provides the endpoints to search, browse, and get various details relating to the book. `BookInfoService` is used to get all the information pertaining to the book. `CatalogueInfoService` also uses the `BookInfoCache` service to fetch data that is already in cache.

- **BookInfoService deployment**: This deployment has a serverless implementation service that fetches various book information from the Google Books API. This will be implemented using the fn project serverless framework running on GraalVM.

- **BookInfoCacheService deployment**: This deployment is a Redis cache that caches all the book information, so as to avoid redundant calls to the Google Books API.

The final finished source code can be found in the Git repository. We will not be discussing the source code, but simply to gain a good understanding of how to build these microservices, we will pick `BookInfoService` and implement it with various microservices frameworks in the next section.

Exploring modern microservices frameworks

There are modern frameworks that are built around creating microservices rapidly. These frameworks are built on the basis of the **Container-First** and **Cloud-First** design principles. They are built from the ground up, with a fast boot time and a low memory footprint. Helidon, Micronaut, and Quarkus are three of the most widely used modern Java frameworks. All three frameworks run natively on GraalVM. Each of these frameworks promises faster startup and a low memory footprint, and they achieve this by means of different methods. Let's explore these frameworks in this section.

To understand these frameworks, let's now get hands-on in building a simple book information service. It is a simple service that accepts a keyword, uses the Google Books API to retrieve the book information, and returns detailed information relating to all the books that match the keyword. The response is returned as **JSON** (**JavaScript Object Notation** – refer to `https://www.json.org/json-en.html` for more details).

Let's first start with a traditional microservice that we build using Spring Boot without GraalVM

Building BookInfoService using Spring without GraalVM

Spring is one of the most widely used Java microservices frameworks. It comes with a lot of great features and is one of the popular frameworks used to build cloud-native applications. In this section, we will build in the traditional way without GraalVM, so as to understand the shortcomings of the traditional approach.

Creating Spring boilerplate code

To create the Spring boilerplate code, let's go to `https://start.spring.io/` on a browser. This website helps us to specify some configurations and generate the boilerplate code. Let's generate the boilerplate code for our `BookInfoService`. The following screenshot shows the Spring initializer:

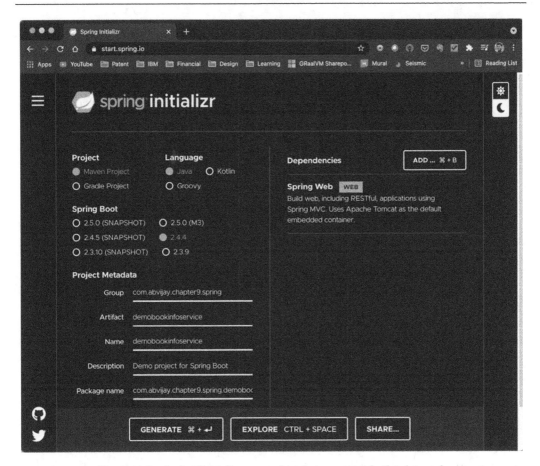

Figure 10.5 – Spring Initializr screenshot that generates boilerplate code

The preceding screenshot shows the configurations selected to generate the boilerplate. To keep it simple and focused, we are selecting **Spring Web**. We will use a simple `HttpClient` to call the Google APIs so as to keep it simple for ou, instead of the recommended way of using jsonb and so on.

We need to extract the ZIP file that is generated and then implement the service. The following is the core logic code snippet. The full code is available in the Git repository at `https://github.com/PacktPublishing/Optimizing-Application-Performance-with-GraalVM`:

```
@RestController
public class BookInfoServiceController {
    @RequestMapping("/book-info")
    public String bookInfo(@RequestParam String query) {
```

In the preceding code, we are setting the path as /book-info in order to call BookInfoService. In the following code, we will call the Google API to get the book information:

```
String responseJson ="{}";
try {
    String url = "https://www.googleapis.com/
        books/v1/volumes?q="+query+"&key=<your google
        api key>";
    HttpClient client = HttpClient.newHttpClient();
    HttpRequest request = HttpRequest.newBuilder()
        .uri(URI.create(url)).build();
    HttpResponse<String> response;
    response =
        client.send(request, BodyHandlers.ofString());
    responseJson = response.body();
} catch (Exception e) {
    responseJson = "{'error','"+e.getMessage()+"'}";
    e.printStackTrace();
}
    return responseJson;
}
}
```

In the preceding code, we are calling the Google Books API using our Google API key. You have to get your own key and include it in the URL. Refer to https://cloud.google.com/apis/docs/overview for more details on how to get your own Google API. We are calling the Google Books API using HttpClient and passing the response to the requester.

Let's now build this code and run it. We will be using Maven to build it. The following command will build the code:

```
./mvnw package
```

This will download all the dependencies, build the application, and generate a JAR file. You will find the JAR file under the target folder. We should be able to run the JAR file using the following command:

```
java -jar target/book-info-service-0.0.1-SNAPSHOT.jar
```

This will start the Spring Boot application. The following screenshot shows the output of running the application:

Figure 10.6 – Output screenshot of the Spring BookInfoService application

Now, let's access this application using a REST client. In this case, we are using `CocoaRestClient` (https://mmattozzi.github.io/cocoa-rest-client/). You can use any REST client or even use the browser to invoke the service. Let's invoke `http://localhost:8080/book-info?query=graalvm`. The following screenshot shows the output:

Figure 10.7 – Output of invoking the BookInformationService Spring application

Now that we know that the application is running, let's package this application into a Docker container and build the image. The following is the Dockerfile code for building the image:

```
FROM adoptopenjdk/openjdk11:ubi
ARG JAR_FILE=target/*.jar
COPY ${JAR_FILE} bookinfo.jar
ENTRYPOINT ["java","-jar","/bookinfo.jar"]
```

This is a very simple Dockerfile. We are building the image using `openjdk11` as the base. We are then copying the jar file that we generated and specifying the entry point to run the jar file when we start the container. Let's now build the Docker image using the following command:

```
docker build -t abvijaykumar/bookinfo-traditional .
```

Please feel free to use your name tag for the Docker image. These Docker images are also available on the author's Docker Hub at `https://hub.docker.com/u/abvijaykumar`. This will build an image. We should be able to see whether the image has been created by using the following command:

```
docker images
```

Let's run this image using the following command:

```
docker run -p 8080:8080 abvijaykumar/bookinfo-traditional
```

The following screenshot shows the output of running the previous command:

Figure 10.8 – Console output of running the BookInformationService Spring application

We can see that it booted up in 2.107 seconds. We should be able to call the service. The following screenshot shows the output after calling `http://localhost:8080/book-info?query=graalvm`:

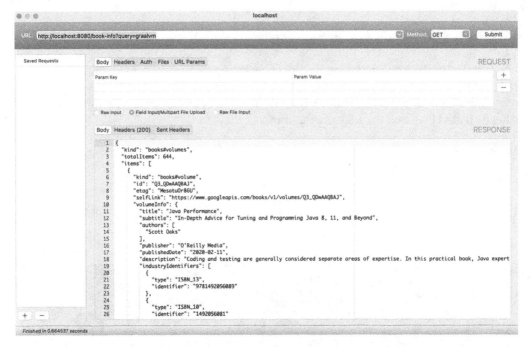

Figure 10.9 – Result of invoking the BookInformationService Spring application in a container

Let's now use the modern frameworks to build the same service to understand and compare how these modern frameworks perform better with GraalVM.

Building BookInfoService with Micronaut

Micronaut is a full-stack microservices framework introduced by the developers of the Grails framework. It has an integration with all the ecosystem and tools and relies on compile-time integration, rather than runtime integration. This makes the final applications run faster, as they are compiled with all the dependencies during build time. It achieves this with annotation and aspect-oriented programming concepts of code injection at build time. This was introduced in 2018. For more details on Micronaut, please refer to `https://micronaut.io/`.

Let's build `BookInfoService` with Micronaut. To get started, we need to install the Micronaut command line. Refer to `https://micronaut.io/download/` for detailed instructions on installing the Micronaut CLI. Once installed, we should be able to call the `mn` command. Let's now create our `BookInfoService` Micronaut boilerplate code with `mn`. The following command creates the boilerplate code. We are passing the `-b=maven` flag to create the Maven build:

```
mn create-app com.abvijay.f.mn.bookinfoservice -b=maven
| Application created at /Users/vijaykumarab/Google Drive/
GraalVM-Book/Code/chapter9/mn/bookinfoservice
```

We should see a directory created called `bookinfoservice` where all the generated boilerplate code is created. Let's now set the environment to point to GraalVM. To validate whether we are using the right version of GraalVM, we can check by running `java-version`. The following output shows the version of GraalVM:

```
java -version
java version "11.0.10" 2021-01-19 LTS
Java(TM) SE Runtime Environment GraalVM EE 21.0.0.2 (build
11.0.10+8-LTS-jvmci-21.0-b06)
Java HotSpot(TM) 64-Bit Server VM GraalVM EE 21.0.0.2 (build
11.0.10+8-LTS-jvmci-21.0-b06, mixed mode, sharing)
```

Let's now update the Micronaut code to implement our logic. The following code snippet shows the code for `Controller`, which exposes the REST endpoint:

```
@Controller("/bookinfo")
public class BookInfoController {
    @Get("get-info")
    public String getBookInfo(String query) {
        BookInfoService svc = new BookInfoService();
        String ret = svc.fetch(query);
        return ret;
    }
}
```

The `BookInfoService` class has the exact same code as what we implemented in Spring Boot in the preceding code. Let's now compile the Micronaut project by executing the following command:

```
./mvnw package
```

We can then run the Micronaut application by executing the following command:

```
./mvnw mn:run
```

The following screenshot shows the output when we run the Micronaut application:

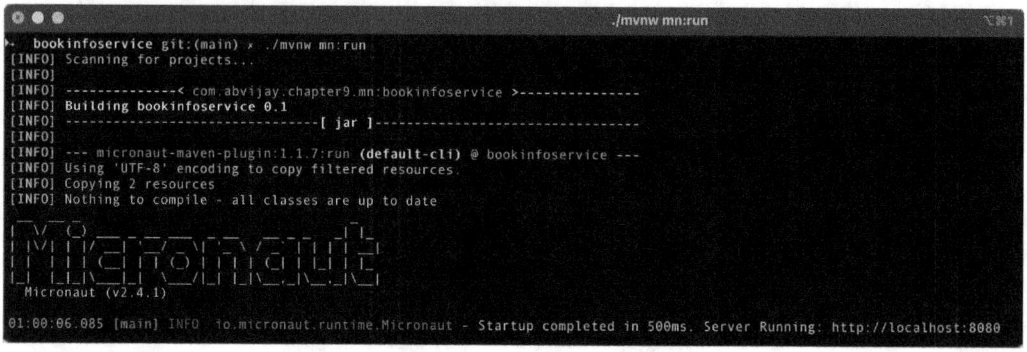

Figure 10.10 – Console output of running the Micronaut BookInformationService Spring application

We can see that it took just 500 milliseconds to load the Micronaut application, as compared to Spring Boot in the *Building BookInfoService using Spring without GraalVM* section, which took around 2 seconds. This is significantly fast, considering how simple and small our application is. Let's now build a Docker image of this application. Micronaut provides a direct way to build a Docker image with Maven by passing the `-Dpackaging=docker` argument. The following command will generate the Docker image directly:

```
mvn package -Dpackaging=docker
```

Micronaut can also generate the Dockerfile so that we can customize and execute separately. The Dockerfiles are created under the target directory when we pass the `-mn:dockerfile` argument to the command. The following is the Dockerfile that is created:

```
FROM openjdk:15-alpine
WORKDIR /home/app
COPY classes /home/app/classes
COPY dependency/* /home/app/libs/
```

```
EXPOSE 8080
ENTRYPOINT ["java", "-cp", "/home/app/libs/*:/home/app/
classes/", "com.abvijay.chapter9.mn.Application"]
```

We can see that the Docker image is built on openjdk. We are still not using the GraalVM native image feature. Let's build this image by calling the following command:

```
docker build -t abvijaykumar/bookinfo-micronaut .
docker images
```

Let's now run this Docker image by calling the following command:=

```
docker run -p 8080:8080 abvijaykumar/bookinfo-micronaut
```

The following shows the output of running the preceding command:

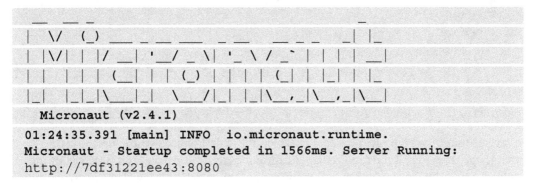

```
  Micronaut (v2.4.1)
01:24:35.391 [main] INFO  io.micronaut.runtime.
Micronaut - Startup completed in 1566ms. Server Running:
http://7df31221ee43:8080
```

We can see that the application started in 1.5 seconds, which is still faster than the Spring image. We are still not using the GraalVM native image feature. Let's now build the same application as a GraalVM native image. To build a native image, Micronaut supports a Maven profile, which can be invoked by passing the -Dpackaging=native-image argument to the command. The following command creates the native image:

```
./mvnw package -Dpackaging=native-image
docker images
```

Let's now generate the Dockerfile to understand how this image is created. To generate the Dockerfile, we need to execute the following command:

```
mvn mn:dockerfile -Dpackaging=docker-native
```

This will generate the Dockerfile under the target directory. The following code shows the Dockerfile:

```
FROM ghcr.io/graalvm/graalvm-ce:java11-21.0.0.2 AS builder
RUN gu install native-image
WORKDIR /home/app
COPY classes /home/app/classes
COPY dependency/* /home/app/libs/
RUN native-image -H:Class=com.abvijay.chapter9.mn.Application
-H:Name=application --no-fallback -cp "/home/app/libs/*:/home/
app/classes/"
FROM frolvlad/alpine-glibc:alpine-3.12
RUN apk update andand apk add libstdc++
COPY --from=builder /home/app/application /app/application
EXPOSE 8080
ENTRYPOINT ["/app/application"]
```

We can see that this is a multi-phase Dockerfile. In the first phase, we are installing the native image, copying all the required application files into the image, and finally running the native-image command to create the native image. In the second phase, we are copying the native image and providing an entry point.

Let's run this image and see how fast it loads. Let's execute the following command:

```
docker run -p 8080:8080 bookinfoservice
```

The following output shows that it took just 551 milliseconds for the image to load, which is almost half the time it took for the non-GraalVM Micronaut application:

```
/app/application: /usr/lib/libstdc++.so.6: no version
information available (required by /app/application)
```

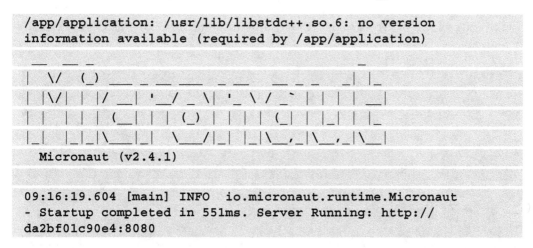

```
   Micronaut (v2.4.1)
```

```
09:16:19.604 [main] INFO  io.micronaut.runtime.Micronaut
- Startup completed in 551ms. Server Running: http://
da2bf01c90e4:8080
```

We can see how easy it is to create a microservice with Micronaut, and how it seamlessly integrates with the GraalVM toolchain to generate Docker images with a very small footprint and fast loading.

Quarkus is another very popular microservices framework. Let's now explore Quarkus and build the same service using Quarkus.

Building BookInfoService with Quarkus

Quarkus was developed by Red Hat and provides the most sophisticated list of integration with the Java ecosystem of frameworks. It is built on top of the MicroProfile, Vert.x, Netty, and Hibernate standards. It is built as a fully Kubernetes-native framework. This was introduced in 2019.

Let's now build `BookInfoService` using Quarkus. Quarkus provides a starter code generator at `http://code.quarkus.io`. Let's go to that website and generate our code. The following screenshot shows the configurations that are selected to generate our `BookInfoService` boilerplate code. We are also including RESTEasy JAX-RS to create our endpoint:

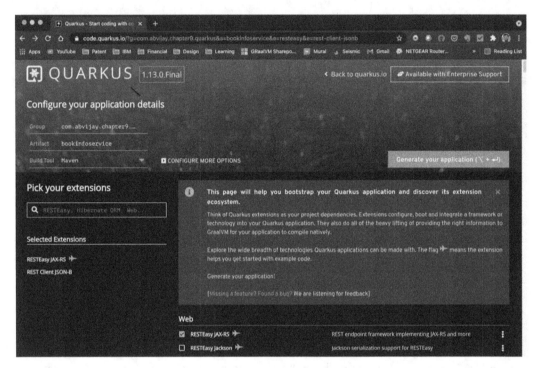

Figure 10.11 – Screenshot of code.quarkus.io to generate boilerplate code

This will generate code in a zip file (we can also provide a Git repository, where Quarkus will automatically push the code). Let's now download the zip file and then extract and compile it with the following command:

```
./mvnw compile quarkus:dev.
```

The best part about Quarkus is that when we execute this command, it provides a way for us to edit the code and test it without restarting the server. This helps in building the application rapidly. Now, let's update the Quarkus code to our `BookInfoService` endpoint.

The following code shows implementation of the endpoint:

```
@Path("/bookinfo")
public class BookInfoService {
    @GET
    @Produces(MediaType.TEXT_PLAIN)
    @Path("/getBookInfo/{query}")
    public String getBookInfo(@PathParam String query) {
        String responseJson = "{}";
        try {
            String url = "https://www.googleapis.com/books/
                v1/volumes?q=" + query
                    + "andkey=<your google api key>";
            HttpClient client = HttpClient.newHttpClient();
            HttpRequest request =
                HttpRequest.newBuilder()
                    .uri(URI.create(url)).build();
            HttpResponse<String> response;
            response = client.send(request,
                BodyHandlers.ofString());
            responseJson = response.body();
        } catch (Exception e) {
            responseJson =
                "{'error', '" + e.getMessage() + "'}";
            e.printStackTrace();
        }
        return responseJson;
    }
}
```

While we update the code and save it, Quarkus automatically updates the runtime. We don't have to restart the server. The following screenshot shows the output of calling our `bookservice` that is running with Quarkus:

Figure 10.12 – Result of invoking the Quarkus implementation of the
BookInformationService application

Let's now build a GraalVM native image using Quarkus. To do that, we need to edit the `pom.xml` file and make sure we have the following profile:

```
<profiles>
    <profile>
        <id>native</id>
        <properties>
            <quarkus.package.type>native</quarkus.package.type>
        </properties>
    </profile>
</profiles>
```

Quarkus uses Mandrel, which is a downstream distribution of GraalVM. You can read more about Mandrel at `https://developers.redhat.com/blog/2020/06/05/mandrel-a-community-distribution-of-graalvm-for-the-red-hat-build-of-quarkus/`.

Let's now build the native image. Quarkus provides a direct Maven profile to build native images. We can create a native image by executing the following command:

```
mvn package -Pnative
```

This will create the native build under the target folder. Let's run the native build directly. The following shows the output after running the native image:

```
./bookinfoservice-1.0.0-SNAPSHOT-runner

 __  ____  __  _____   ___  __ ____  _____
 --/ __ \/ / / / _ | / _ \/ //_/ / / / __/
 -/ /_/ / /_/ / __ |/ , _/ ,< / /_/ /\ \
 --_____/_/ |_/_/|_/_/|_|\____/___/
2021-03-31 11:26:31,564 INFO  [io.quarkus] (main)
bookinfoservice 1.0.0-SNAPSHOT native (powered by
Quarkus 1.13.0.Final) started in 0.015s. Listening on:
http://0.0.0.0:8080
2021-03-31 11:26:31,843 INFO  [io.quarkus] (main) Profile prod
activated.
2021-03-31 11:26:31,843 INFO  [io.quarkus] (main) Installed
features: [cdi, rest-client, rest-client-jsonb, resteasy]
```

We can see that it took just `0.015s` to start the application. This is significantly faster than the traditional implementations, which took around 2 seconds to start up.

Quarkus also created various Dockerfile versions, and we can find these under the Docker folder. The following screenshot shows the list of Dockerfiles that Quarkus automatically creates:

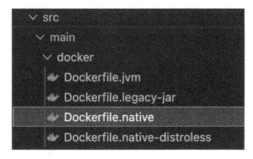

Figure 10.13 – Screenshot of various versions of Dockerfiles that Quarkus creates

Let's quickly explore these various types of Dockerfiles:

- **Dockerfile.legacy-jar** and **Dockerfile.jvm**: This Dockerfile has the commands to build a Docker image with a normal Quarkus application, JAR, and OpenJDK headless.

- **Dockerfile.native**: This Dockerfile builds the native image.

- **Dockerfile.native-distroless**: This Docker file also generates an image with a native image, but uses the new technique introduced by Google to build the image that contains just the application, language runtime, and no operating system distribution. This helps in creating a small image, and has fewer vulnerabilities. Refer to `https://github.com/GoogleContainerTools/ distroless` for more details on distroless containers.

We can create Docker images of these various Docker versions by executing the following commands:

```
docker build -f ./src/main/docker/Dockerfile.jvm -t
abvijaykumar/bookinfo-quarkus-jvm .
```

```
docker build -f ./src/main/docker/Dockerfile.jvm-legacy -t
abvijaykumar/bookinfo-quarkus-jvm-legacy .
```

```
docker build -f ./src/main/docker/Dockerfile.native -t
abvijaykumar/bookinfo-quarkus-native .
```

```
docker build -f ./src/main/docker/Dockerfile.native-distroless
-t abvijaykumar/bookinfo-quarkus-native-distroless .
```

To compare the sizes of these images, lets run the following command:

```
docker images
```

The following chart compares the sizes of each of these images:

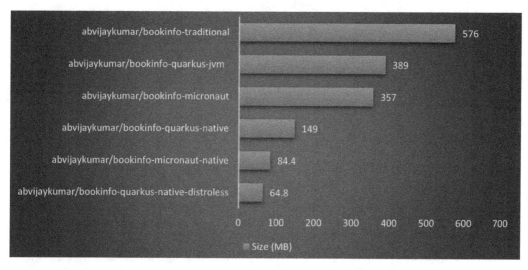

Figure 10.14 – Chart comparing the size of Docker images

At the time of writing this book, the smallest footprint and the fastest executing GraalVM microservice image is built using a Quarkus native distroless image. Spring has also launched Spring Native (`https://spring.io/blog/2021/03/11/announcing-spring-native-beta`) and Oracle has Helidon (`https://helidon.io/#/`), which provide similar frameworks to run on GraalVM.

Building a serverless BookInfoService using fn project

Function-as-a-Service, or serverless, is another architectural pattern for running code on demand and utilizes cloud resources. The serverless approach runs the code when a request is received. The code boots up, executes, handles the requests, and then shuts down, thereby utilizing cloud resources to the optimum. This provides a highly available, scalable architecture at optimum cost. However, serverless architecture demands a faster boot, quicker execution, and shutdown.

GraalVM native images (ahead of time) is the best option for serverless, as native images start up and run faster than traditional Java applications. GraalVM native images have a very small footprint, they are fast to boot, and they come with embedded VM (Substrate VM).

fn project is also a great environment for building serverless applications. Fn supports the building of serverless applications in Go, Java, JavaScript, Python, Ruby, and C#. It is a very simple and rapid application development environment that comes with an fn daemon and a CLI that provides most of the scaffolding to build serverless applications.

In this section, let's focus on building the `BookInfoService` function using fn project. Please refer to `https://fnproject.io/` for detailed instructions on installing the `fn` command-line interface. We first have to start the fn daemon server using `fn start`. The fn server runs in Docker, and you can check that by running `docker ps`. The Fn daemon server runs at port `8080`.

The `fn` command line also provides a way to generate boilerplate code. Let's now generate the project by executing the following command:

```
fn init --runtime java book-info-service-function
Creating function at: ./book-info-service-function
Function boilerplate generated.
func.yaml created.
```

This creates a `book-info-service-function` directory with all the boilerplate code. Let's inspect what is inside that directory. We will find `func.yaml`, `pom.xml`, and `src` directories.

`func.yml` is the main manifest `yaml` file that has the key information about the class that implements the function and the entry point. Let's inspect the configuration file:

```
schema_version: 20180708
name: book-info-service-function
version: 0.0.1
runtime: java
build_image: fnproject/fn-java-fdk-build:jdk11-1.0.124
run_image: fnproject/fn-java-fdk:jre11-1.0.124
cmd: com.example.fn.HelloFunction::handleRequest
```

Let's now understand the preceding configuration file:

- `name`: The name of the function. We can see the name of the function that we specified in our `fn init` command line.

- `version`: The version of this function.

- `runtime`: JVM as the runtime.

- `build_image`: The Docker image that should be used to build the Java code; in this case, we see that it's JDK 11.

- `run_image`: The Docker image that should be used as a runtime; in this case, it is JRE11.

- cmd: This is the entry point, which is `ClassName:MethodName`. We will change cmd to point to our class and method: `cmd: com.abvijay.chapter9.fn.BookInfoService::getBookInfo`.

In the `src` folder, we will create `com.abvijay.chapter9.fn.BookInfoService` with the `getBookInfo()` method. The implementation of `getBookInfo()` is the same as other implementations we performed previously in the section.

The following code shows the implementation of the function that calls the Google API to get the books:

```java
public String getBookInfo(String query) {
    String responseJson = "{}";
    try {
        String url = "https://www.googleapis.com/
            books/v1/volumes?q=" + query
                + "&key=<your_google_api_key>";
        HttpClient client = HttpClient.newHttpClient();
        HttpRequest request = HttpRequest.newBuilder()
            .uri(URI.create(url)).build();
        HttpResponse<String> response;
        response = client.send(request,
            BodyHandlers.ofString());
        responseJson = response.body();
    } catch (Exception e) {
        responseJson =
            "{'error', '" + e.getMessage() + "'}";
        e.printStackTrace();
    }
    return responseJson;
}
```

Let's now build and deploy this serverless container to the local Docker. Functions are grouped into applications. An application can have multiple functions. This helps in grouping and managing them. So we need to create a book info service app using the `fn create app` command. The following shows the output after executing the command:

```
fn create app book-info-service-app
Successfully created app:  book-info-service-app
```

Once the app has been created, we can deploy it using the `fn deploy` command. This command has to be executed at the root folder of the function app that we created. The following shows the output after executing the command:

```
fn deploy --app book-info-service-app --local
Deploying book-info-service-function to app: book-info-service-app
Bumped to version 0.0.2
Building image book-info-service-function:0.0.2 ..............
.................................................................
.............................
Updating function book-info-service-function using image book-info-service-function:0.0.2...
Successfully created function: book-info-service-function with book-info-service-function:0.0.2
```

The `fn deploy` command will build the code using Maven, package it as a Docker image, and deploy it to the local Docker runtime. fn can also be used to deploy to the cloud or k8s cluster directly.

Let's now use the `docker images` command to check whether our image has been built:

```
docker images
```

We can also use `fn inspect` to get all the details about the function. This helps in discovering the services. The following shows the output of executing the command:

```
fn inspect function book-info-service-app book-info-service-function
{
    "annotations": {
        "fnproject.io/fn/invokeEndpoint":
        "http://localhost:8080/invoke/
        01F29E8SXKNG8G00GZJ0000002"
    },
    "app_id": "01F29E183WNG8G00GZJ0000001",
    "created_at": "2021-04-02T14:02:25.587Z",
    "id": "01F29E8SXKNG8G00GZJ0000002",
    "idle_timeout": 30,
    "image": "book-info-service-function:0.0.2",
```

```
    "memory": 128,
    "name": "book-info-service-function",
    "timeout": 30,
    "updated_at": "2021-04-02T14:02:25.587Z"
}
```

Now let's invoke the service. Since our function expects an input argument in the number, we can pass it using an echo command and pipe the output to `fn invoke` to invoke our function:

```
echo -n 'java' | fn invoke book-info-service-app book-info-
service-function
{
  "kind": "books#volumes",
  "totalItems": 1941,
  "items": [
    {
        "kind": "books#volume",
        "id": "Q3_QDwAAQBAJ",
        "etag": "OHl3HzInpzY",
        "selfLink": "https://www.googleapis.com/
            books/v1/volumes/Q3_QDwAAQBAJ",
                "volumeInfo": {
        "title": "Java Performance",
...
```

We can see the function executing and the output of the Google API (the preceding output is partial, to save space). Now let's run the same logic on GraalVM.

The base image for GraalVM is different. We use `fnproject/fn-java-native-init` as the base, and initialize our fn project with that. The following is the output of generating a Graal native image-based fn project:

```
fn init --init-image fnproject/fn-java-native-init book-info-
service-function-graal
Creating function at: ./book-info-service-function-graal
Running init-image: fnproject/fn-java-native-init
Executing docker command: run --rm -e FN_FUNCTION_NAME=book-
info-service-function-graal fnproject/fn-java-native-init
func.yaml created.
```

This generates a Dockerfile. This is followed by the Dockerfile code, which you can find under the project directory (`book-info-service-function-graal`). This `fn` configuration works differently. It also generates a Dockerfile, with all the necessary Docker build commands. This is a multi-stage Docker build file. Let's inspect this Dockerfile:

```
17   FROM fnproject/fn-java-fdk-build:1.0.124 as build
18   WORKDIR /function
19   ENV MAVEN_OPTS=-Dmaven.repo.local=/usr/share/maven/ref/repository
20   ADD pom.xml pom.xml
21   RUN ["mvn", "package", "dependency:copy-dependencies", "-DincludeScope=runtime", "-DskipTests=true", "-Dmdep.prependGroupId=true",
22   ADD src src
23   RUN ["mvn", "package"]
24
25   FROM fnproject/fn-java-native:1.0.124 as build-native-image
26   WORKDIR /function
27   COPY --from=build /function/target/*.jar target/
28
29 ∨ RUN /usr/bin/native-image \
30        --static \
31        --no-fallback \
32        --allow-incomplete-classpath \
33        --enable-all-security-services \
34        --enable-url-protocols=https,http \
35        --report-unsupported-elements-at-runtime \
36        -H:Name=func \
37        -classpath "target/*"\
38        com.fnproject.fn.runtime.EntryPoint
39
40
41   FROM busybox:glibc
42   WORKDIR /function
43   COPY --from=build-native-image /function/func func
44   COPY --from=build-native-image /function/runtime/lib/* .
45   ENTRYPOINT ["./func", "-XX:MaximumHeapSizePercent=80"]
46   CMD [ "com.example.fn.Book-info-service-function-graal::handleRequest" ]
47
```

Figure 10.15 – Dockerfile generated by fn

Let's understand this Dockerfile:

- **Line 17**: The image will be built using `fnproject/fn-java-fdk-build`.

- **Line 18**: This sets the working directory to `/function`.

- **Lines 19–23**: Then, the Maven environment is configured.

- **Lines 25–40**: Using `fnproject/fn-java-native` as the base image, the GraalVM is configured and the fn runtime is compiled. This is a very important step as this is what makes our serverless runtime faster and with a smaller footprint.

- **Lines 43–47**: The native images are copied using the `busybox:glibc` (which is the minimal version of Linux+glibc) base image.

- **Line 48**: This is the function entry point. `func.yml`, in this way of building the serverless image, has no information. fn will use the Dockerfile to perform the build (along with Maven) and deploy the image to the repository.

We need to change line 48 to point to our class. Let's replace that with the following:

```
CMD ["com.abvijay.chapter9.
fn.BookInfoService::BookInfoService"]
```

Another important configuration file that we need to change is reflection. json under src/main/conf. This JSON file has the manifest information about the class name and the method. It is used by the native image builder to resolve the reflection we do by calling our function dynamically. Refer to the *Building native images* section in *Chapter 5*, *Graal Ahead-of-Time Compiler and Native Image*.

Now, let's create an fn app and deploy this app using the fn create app command. The following is the output after executing the command:

```
fn create app book-info-service-app-graal
  Successfully created app:  book-info-service-app-graal
```

We can build the native image and deploy it using the fn deploy -app book-info-service-app-graal command and we can execute the method by calling echo -n 'java' | fn invoke book-info-service-app-graal book-info-service-function. Checking the Docker images, we will see that the size of the Java image is 238 MB, and that of the GraalVM image is just 41 MB. That is a 10-times-smaller footprint than traditional Java applications. We can time the function calls, and we can see that the native images are much faster (up to 30%).

```
docker images
```

Serverless is the best solution, for quicker and stateless services, as it does not take any resources and we don't have to keep it running all the time.

In this section, we have looked at various framework implementations and ways to optimize the image.

Summary

Congratulations on reaching this point! In this chapter, we looked at how microservices architectures are built. To understand the architectural thought process, we picked a simple case study and explored how it can be deployed on Kubernetes as a collection of microservices. We then explored various microservices frameworks, and built a service on each of these frameworks, to appreciate the benefits that GraalVM brings to cloud-native architectures.

After reading this chapter, you should have acquired a good understanding of how to go about building microservices-based, cloud-native applications using GraalVM as the runtime. This chapter gives a good head start for Java developers to quickly start building applications on one of the microservices frameworks (Quarkus, Spring, Micronaut). The source code that is provided along with this chapter (in Git) will also provide a good reference implementation of microservices on GraalVM.

Questions

1. What is a microservice?

2. What are the advantages of a microservices architecture?

3. Why is GraalVM an ideal application runtime for microservices?

Further reading

- Microservices architecture (`https://microservices.io/`)

- Micronaut (`http://microprofile.io/`)

- Quarkus (`https://quarkus.io/`)

- Spring Boot (`https://spring.io/`)

- Spring Native (`https://docs.spring.io/spring-native/docs/current/reference/htmlsingle/`)

Assessments

This section contains answers to the questions from all chapters.

Chapter 1 – Evolution of Java Virtual Machine

1. Java code is compiled to bytecode. JVM uses interpreters to convert the bytecode to machine language and uses JIT compilers to compile the most commonly used code snippets (hotspots). This approach helps Java to achieve "write-once run-anywhere," as a result of which programmers don't have to write machine-specific code.

2. A class loader subsystem is responsible for loading the classes. It not only finds the classes, but also verifies and resolves the classes.

3. JVM has five memory areas:

   a. Method: A shared area, where all the class-level data is stored at the JVM level

   b. Heap: All instance variables and objects stored at the JVM level (shared across threads)

   c. Stack: A runtime stack per thread to store the local variables at the method scope, as well as operands and frame data

   d. Registries: PC registers with the addresses of current executing instructions (for each thread)

   e. Native method stack: Native method information for each thread that is used to invoke native methods

Chapter 2 – JIT, Hotspot, and GraalJIT

1. A code cache is a special memory area within JVM that is used by JVM to store the compiled code. The code is compiled by JIT compilers and stored in the code cache. If a method is compiled and is found in the code cache, JVM will use that code to run, instead of interpreting the method code. Refer to the *Code cache* section for more details

2. The code cache size can be changed using the following flags for fine-tuning. Refer to the *Code cache* section for more details:

    a. `-XX:InitialCodeCacheSize` – The initial size of the code cache. The default size is 160 KB (the size varies based on the JVM version)

    b. `-XX:ReservedCodeCacheSize` – This is the maximum size the code cache can grow to. The default size is 32/48 MB. When the code cache reaches this limit, JVM will throw a warning, `CodeCache is full. Compiler has been disabled`. JVM offers the `UseCodeCacheFlushing` option to flush the code cache when the code cache is full. The code cache is also flushed when the compiled code is not hot enough (when the counter is less than the compiler threshold).

    c. `-XX:CodeCacheExpansionSize` – This is the expansion size. When it scales up, its default value is 32/64 KB.

3. The compiler threshold is the factor that is used to decide when the code is "hot." When the code reaches the compiler threshold, JVM will spin off the JIT compilation (C1 or C2) on a compilation thread. Refer to the *Compiler threshold* section for more details.

4. Sometimes, the code can become hot while it is running a long-running loop. In such cases, JVM will compile that code and perform OSR. Refer to the *On-stack replacement* section for more details and a detailed flow chart regarding how JVM performs OSR.

5. In JVM, there is an interpreter and two types of compiler – C1 and C2. Users can specify any specific compiler to be used to optimize the code. By default, JVM performs tiered compilation, which is a combination of C1 and C2, based on various compiler thresholds. There are five tiers:

    a. Interpreted code (level 0)

    b. Simple C1 compiled code (level 1)

    c. Limited C1 compiled code (level 2)

d. Full C1 compiled code (level 3)

e. C2 compiled code (level 4)

There are three main patterns that JVM follows:

a. Normal Flow

b. C2 Busy

c. Trivial Code

6. Refer to the *Tiered compilation* section for more details.

7. Inlining is one of the key optimization techniques that JIT compilers use. Based on the profiling of the code, JIT identifies the methods that can be inlined in order to avoid method calls. Method calls are expensive, as it performs jumps and stack frames are created.

8. Monomorphic dispatch is another optimization technique used to identify the specific implementations of a polymorphic implementation. JIT profiles the code, identifies the specific implementation, and optimizes the code around that. Please refer to the *Monomorphic, bimorphic, and megamorphic dispatch* section for more details.

9. Loop unrolling is one of the most effective optimizations that JIT performs, by inlining code in the loop body, with additional code, and reducing the number of iterations a loop has to iterate through. Please refer to the *Loop optimization – loop unrolling* section for more details and examples.

10. Escape analysis is an optimization technique that the JIT profiler performs to identify the allocation and scope of the variables, and takes decisions in avoiding heap allocation, and replaces that with stack allocation, based on the scope of the variable. This is one of the most advanced analyses performed by JIT profilers. Please refer to the *Escape analysis* section for more details.

11. JIT performs deoptimization when any of the optimistic assumptions that were made to optimize and compile the code are invalid. JIT will make the compiled code non-entrant and fall back to the interpreter.

12. JVMCI stands for *Java Virtual Machine Compiler Interface*. This interface was added to the JDK in Java 9. JVMCI provides an API to extend JVM and build custom compilers. Graal JIT is an implementation of JVMCI. Please refer to the *Graal JIT and the JVM Compiler Interface (JVMCI)* section for more details.

Chapter 3 – Graal VM Architecture

1. GraalVM comes in two versions – Community Edition and Enterprise Edition. Refer to *Reviewing the GraalVM editions* section for more details.

2. JVMCI stands for *Java Virtual Machine Compiler Interface*. Java 9 and above provide a way to implement custom JIT compilers. JVMCI provides an API to implement these custom compilers and provides access to JVM objects and the code cache. Graal JIT is an implementation of JVMCI. Refer to the *Java Virtual Machine Compiler Interface (JVMCI)* section for more details.

3. Graal JIT replaces the C2 JIT compiler. Graal JIT is completely written in Java from the ground up but uses the hardened logic and best practices of the C2 compiler. Graal JIT implements better optimization strategies than C2 JIT, making it the best JIT compiler for Java. Graal JIT can also be used to compile other languages that are converted into intermediate representations in order to use the advanced optimization strategies. Refer to the *Graal compiler and tooling* section for more details.

4. Graal JIT requires a considerable amount of time to warm up, profile, and optimize the code. In certain use cases, this may not be suitable (such as serverless or containers). For such cases, Graal provides AOT compilation to compile the code directly to the native image.

5. Graal AOT optimization is more related to static code analysis, but it does now have the runtime profile of the code to apply any advanced optimization. **Profile Guided Optimization** (**PGO**) provides a way to compile the code with instrumentation, generate a profile of the runtime, and use that profile to recompile the code to the most optimum native image. Refer to the *SubstrateVM (Graal AOT, Native Image)* section for more details.

6. The Truffle framework is built on top of Graal to support non-JVM languages to run on Graal JVM. Truffle provides the Truffle Language Implementation API and various other polyglot APIs to provide a very sophisticated environment where code in multiple languages can be embedded and interact. Refer to the *Truffle* section for more details.

7. SubstrateVM is an embeddable VM that can be packaged along with the native images that are compiled by the Graal AOT compiler. Refer to the *SubstrateVM (Graal AOT and Native Image)* section for more details.

8. Guest Access Context is an object that is used by the host language (such as Java) to provide access to the guest language (such as JavaScript) to various OS resources, such as the filesystem, I/O, and thread. Refer to the *Security* section for more details.

9. GraalVM provides the most advanced JIT compilation, ideal for long-running processes involving high throughput. The GraalVM AOT compiler, along with SubstrateVM, provides the smallest and fastest runtime for cloud-native microservices implementations. Combined with PGO, it generates the optimum code to run on the cloud. Refer to the *GraalVM microservices architecture overview* section for more details.

Chapter 4 – Graal Just-In-Time Compiler

1. Graal JIT compilation can be divided into two phases: frontend and backend.

    The frontend phase is platform-independent compilation, where the code is converted to a platform-independent intermediate representation called **High-Level Intermediate Representation (HIR)**, represented via Graal Graphs. This HIR is optimized in three tiers: High, Medium, and Low.

    The backend phase is more platform-dependent compilation, where a **Low-Level Intermediate Representation (LIR)** is created and optimized at the machine code level. These optimizations are platform dependent.

    Refer to the *Graal JIT compilation pipeline and Tiered Optimization* section for more details.

2. **Intermediate Representations (IRs)** are among the most important data structures for compiler design. IRs provide a graph that helps the compiler understand the structure of the code, identify opportunities, and perform optimizations. Refer to the *Graal Intermediate Representation* section for more details.

3. **Static Single Assignment (SSA)** is a form used in IRs where each variable is assigned once, and any time there is a change in the value, a new variable is used. Every variable is declared before it is used. This helps us to keep track of variables and values and helps optimize the code better using graphs. Refer to the *Graal intermediate representation* section for more details.

4. Speculative optimization is a compiler optimization technique of performing various code optimizations with speculation. Speculations are assumptions that are made based on profiling the code. The optimizations are performed on the code based on these assumptions. When these assumptions are proven wrong, at runtime, a deoptimization is performed. This helps to optimize focused parts of the code, instead of the whole code, which might slow down the runtime. Refer to the *Graal compiler optimizations* section for more details.

5. Escape analysis is an optimization technique that identifies the scope and usage of the objects and decides the allocation of the objects either on the heap or the stack or the register. This has a significant impact on memory usage and performance. Escape analysis is performed at the method level, while partial escape analysis performs a deeper analysis of the code to track the objects not just at the method-level scope, but also at the control block level. This helps further optimize the code. Refer to the *Partial escape analysis* section for more details.

Chapter 5 – Graal Ahead-of-Time Compiler and Native Image

1. GraalVM comes with a tool called the Native Image builder, `native-image`. This can be used to compile ahead of time and create a native image. Please refer to the *Building native images* section for more details.

2. The Native Image builder, when it compiles the code ahead of time, performs points-to analysis to understand all the dependent classes and methods that are accessed by the application code. It uses this information to optimize the native image, by only building the required code into the image. This provides faster execution and smaller images. Please refer to the *Building native images* section for more details.

3. The Native Image builder performs region analysis to initialize classes ahead of time into the heap so that the startup of the native image is faster. Please refer to the *Building native images* section for more details.

4. The Native Image builder packages the **Garbage Collector** (**GC**) code along with the native image. There are two types of GC that can be enabled in the native image. The Serial GC is a default GC and is available both in the Community and Enterprise editions. G1 performs more advanced garbage collection and is only available in the Enterprise edition. Please refer to the *Native Image memory management configurations* section for more details.

5. The Native Image builder can only perform static code analysis, unlike the JIT compiler, which can perform the runtime profiling of the code and optimize the code at runtime. PGO brings the runtime profiling information to a native image for further optimization. Please refer to the *Profile-guided optimization (PGO)* section for more details.

6. Since native images are built ahead of time, the Native Image builder needs to have all the classes loaded at build time. Hence, native images have limitations in supporting dynamic features such as reflection and JNI. However, GraalVM's Native Image builder provides ways to pass dynamic resource information at build time. Please refer to the *Native image configuration* and *Limitations of Graal AOT (Native Image)* sections for more details.

Chapter 6 – Truffle for Multi-language (Polyglot) support

1. Specialization is a key optimization that helps identify the specific type of a variable. In dynamically typed languages, the type of a variable is not declared in the code. The interpreter starts assuming generic types and, based on the runtime profiling, will speculate on the type of the variable. Please refer to the *Truffle interpreter/ compiler pipeline* section for more details.

2. When Truffle speculates on a specialized type of a node, the node is rewritten dynamically, and the Truffle AST provides a way to rewrite the nodes to optimize the AST before submitting it to Graal for further optimized execution. Please refer to the *Truffle interpreter/compiler pipeline* section for more details.

3. When Truffle finds that the AST has not been rewritten, it assumes that the AST has stabilized. The code is then compiled to machine code for the guest language after aggressive constant folding, inlining, and escape analysis. This is called Partial Evaluation. Please refer to *Partial Evaluation* in the *Truffle interpreter/compiler pipeline* section for more details.

4. Truffle provides a Domain-Specific Language implemented as annotation generators. This helps guest language developers write smaller code and focus on the logic, rather than the boilerplate code. Please refer to the *Truffle DSL* section for more details.

5. A frame is a Truffle class that provides the interface to read and store data in the current namespace. Refer to *Frame management and local variables* in the *Truffle interoperability* section for more details.

6. Truffle defines a Dynamic Object Model to provide a standard interface and framework for various guest language implementations to have a standard way of defining and exchanging data. Refer to *Dynamic Object Model* in the *Truffle interoperability* section for more details.

Chapter 7 – GraalVM Polyglot – JavaScript and Node.js

1. `Polyglot` is the object that is used in JavaScript to run other language code. We use the method `eval()` to run the code. Please refer to the *JavaScript interoperability* section for more details on how to use this object to run the code.

2. The `Context` object provides the polyglot context to allow the guest language code to run in the host language. A polyglot context represents the global runtime state of all installed and permitted languages. Please refer to the *JavaScript embedded code in Java* section for more details on how to use this object to run the code.

3. The `Context` object helps provide fine-grained access control. The access control can be controlled with `ContextBuilder`. Please refer to the *JavaScript embedded code in Java* section for more details on how to use this object to run the code.

4. GraalVM provides a Native Image builder option to build native images of applications that have multiple languages embedded. A language flag is used to let the Native Image builder know which languages are used in the application. This flag can also be specified in `native-image` property files. Refer to the *Polyglot native images* section in this chapter to understand more. Refer to *Chapter 5 , Graal Ahead-of-Time Compiler and Native Image* for more details about the native image.

5. The `binding` object acts as an intermediate layer between Java and JavaScript to access methods, variables, and objects between the languages. Please refer to the *Bindings* section to find out more about the binding object and how it is used as an intermediate layer between languages.

Chapter 8 – GraalVM Polyglot – Java on Truffle, Python, and R

1. Java on Truffle is the new way to run Java programs on top of the Truffle framework. Java on Truffle provides an interpreter that is completely built on Java and runs in the same memory space as other Truffle languages. This was introduced in GraalVM version 21. For more details refer to the *Understanding Espresso (Java on Truffle)* section.

2. Java on Truffle provides an isolationist layer, which helps to run untrusted code and code written in an older version of JDK, and provides hot-swap and other advanced features. To learn more about the advantages of using Java on Truffle, refer to the *Why do we need Java on Java?* section for more details.

3. The `Polyglot.cast()` method is used in Java on Truffle to typecast the data that is exported or returned by dynamic languages. Refer to the *Exploring Espresso interoperability with other Truffle languages* section for more details and code examples.

4. **SST** stands for **Simple Syntax Tree** and **ST** stands for **Scope Tree**. Python generates these intermediate representations before converting them into an AST intermediate representation. Python does this using the ANTLR parser and the cache, and speeds up parsing. Refer to the *Understanding Graalpython compilation and interpreter pipeline* section for more details.

5. A `.pyc` file is a cache Python creates after parsing Python code and generating SST and ST representations. This helps speed up parsing the next time the Python module is loaded. Python automatically keeps this cache validated. Refer to the *Understanding Graalpython compilation and interpreter pipeline* section for more details.

6. `polyglot.import_value()` is used to import definitions from other dynamic languages, and `polyglot.export_value()` is used to export Python definitions to other languages. `polyglot.eval()` is used to execute other language code. Refer to the *Exploring interoperability between Python and other dynamic languages* section for more detailed explanations and sample code.

7. In R, we use the `import()` function to import the definitions from other languages. Refer to the *Exploring interoperability of R* section for more details.

8. We use `java.type('classname')` to load a Java class and interoperate with it. This function provides the class, and we can use the `new()` function to create an instance of the object. Refer to the *Exploring the interoperability of R* section for more details and sample code.

Chapter 9 – GraalVM Polyglot – LLVM, Ruby, and WASM

1. Sulong is an LLVM interpreter that is written in Java and internally uses the Truffle language implementation framework. This enables all language compilers that can generate LLVM IR to directly run on GraalVM. Refer to the *Understanding LLVM – the (Sulong) Truffle interface* section for more details.

2. GraalVM Enterprise Edition provides a managed environment of LLVM. The managed mode of execution provides a safe runtime, which, with additional safety, guarantees to catch illegal pointer accesses and access arrays outside of the bounds.

The TruffleRuby interpreter interoperates with the LLVM interpreter to implement the C extensions. This also extends the possibility to use other LLVM languages, such as Rust and Swift, to run as Ruby extensions. Refer to the *Understanding the TruffleRuby interpreter/compiler pipeline* section for more details.

3. WASM is binary code that can run on modern web browsers. It has a very small footprint and performs much faster than JavaScript. Refer to the *Understanding WASM* section for more details.

    Emscripten or `emcc` is the compiler that generates the WASM binary image (`.wasm`) files. Refer to the *Installing and running GraalWasm* section for more details.

Chapter 10 – Microservices Architecture with GraalVM

1. Microservices is an architectural pattern that decomposes a large application into smaller, manageable, and self-contained components that expose the functionality through a standard interface called services. Please refer to the *Microservices architecture overview* section for more details.

2. The microservices architecture pattern helps us build an application that is scalable, manageable, and loosely coupled. This is very important for building cloud-native applications in order to get the most out of the cloud infrastructure and services. Please refer to the *Microservices architecture overview* section for more details.

3. GraalVM provides a high-performance runtime for JVM and non-JVM languages with a small footprint, which is critical for building scalable cloud-native applications. Refer to the *Reviewing modern architectural requirements* section in *Chapter 3, Graal VM Architecture,* and the *Understanding how GraalVM helps build a microservice architecture* section in this chapter for more details.

`Packt.com`

Subscribe to our online digital library for full access to over 7,000 books and videos, as well as industry leading tools to help you plan your personal development and advance your career. For more information, please visit our website.

Why subscribe?

- Spend less time learning and more time coding with practical eBooks and Videos from over 4,000 industry professionals

- Improve your learning with Skill Plans built especially for you

- Get a free eBook or video every month

- Fully searchable for easy access to vital information

- Copy and paste, print, and bookmark content

Did you know that Packt offers eBook versions of every book published, with PDF and ePub files available? You can upgrade to the eBook version at `packt.com` and as a print book customer, you are entitled to a discount on the eBook copy. Get in touch with us at `customercare@packtpub.com` for more details.

At `www.packt.com`, you can also read a collection of free technical articles, sign up for a range of free newsletters, and receive exclusive discounts and offers on Packt books and eBooks.

Other Books You May Enjoy

If you enjoyed this book, you may be interested in these other books by Packt:

Jakarta EE Cookbook

Elder Moraes

ISBN: 978-1-83864-288-4

- Explore Jakarta EE's latest features and API specifications and discover their benefits
- Build and deploy microservices using Jakarta EE 8 and Eclipse MicroProfile
- Build robust RESTful web services for various enterprise scenarios using the JAX-RS, JSON-P, and JSON-B APIs

Microservices with Spring Boot and Spring Cloud, 2nd Edition

Magnus Larsson

ISBN: 978-1-80107-297-7

- Build cloud-native production-ready microservices with this comprehensively updated guide

- Understand the challenges of building large-scale microservice architectures

- Learn how to get the best out of Spring Cloud, Kubernetes, and Istio in combination

Packt is searching for authors like you

If you're interested in becoming an author for Packt, please visit authors. packtpub.com and apply today. We have worked with thousands of developers and tech professionals, just like you, to help them share their insight with the global tech community. You can make a general application, apply for a specific hot topic that we are recruiting an author for, or submit your own idea.

Leave a review - let other readers know what you think

Please share your thoughts on this book with others by leaving a review on the site that you bought it from. If you purchased the book from Amazon, please leave us an honest review on this book's Amazon page. This is vital so that other potential readers can see and use your unbiased opinion to make purchasing decisions, we can understand what our customers think about our products, and our authors can see your feedback on the title that they have worked with Packt to create. It will only take a few minutes of your time, but is valuable to other potential customers, our authors, and Packt. Thank you!

Index

www.ingramcontent.com/pod-product-compliance
Lightning Source LLC
LaVergne TN
LVHW081331050326
832903LV00024B/1121

* 9 7 8 1 8 0 0 5 6 4 9 0 9 *